Reviews of Physiology, Biochemistry and Pharmacology 154

Reviews of

**154 Physiology
Biochemistry and
Pharmacology**

Special Issue on Sensory Systems
Edited by S. Offermanns

Editors
S.G. Amara, Pittsburgh • E. Bamberg, Frankfurt
S. Grinstein, Toronto • S.C. Hebert, New Haven
R. Jahn, Göttingen • W.J. Lederer, Baltimore
R. Lill, Marburg • A. Miyajima, Tokyo
H. Murer, Zurich • S. Offermanns, Heidelberg
G. Schultz, Berlin • M. Schweiger, Berlin

With 16 Figures, 4 in color and 5 Tables

 Springer

ISSN 030-4240
ISBN 978-3-642-06779-2
e-ISBN 978-3-540-32431-7

Springer is a part of Springer Science+Business Media

springer.com

Editor: Simon Rallison, London
Desk Editor: Anne Clauss, Heidelberg

Rev Physiol Biochem Pharmacol (2005) 155:1–35
DOI 10.1007/s10254-004-0038-0

A. Bigiani · C. Mucignat-Caretta · G. Montani · R. Tirindelli

Pheromone reception in mammals

Published online: 31 March 2005

Abstract Pheromonal communication is the most convenient way to transfer information regarding gender and social status in animals of the same species with the holistic goal of sustaining reproduction. This type of information exchange is based on pheromones, molecules often chemically unrelated, that are contained in body fluids like urine, sweat, specialized exocrine glands, and mucous secretions of genitals. So profound is the relevance of pheromones over the evolutionary process that a specific peripheral organ devoted to their recognition, namely the vomeronasal organ of Jacobson, and a related central pathway arose in most vertebrate species. Although the vomeronasal system is well developed in reptiles and amphibians, most mammals strongly rely on pheromonal communication. Humans use pheromones too; evidence on the existence of a specialized organ for their detection, however, is very elusive indeed. In the present review, we will focus our attention on the behavioral, physiological, and molecular aspects of pheromone detection in mammals. We will discuss the responses to pheromonal stimulation in different animal species, emphasizing the complicacy of this type of communication. In the light of the most recent results, we will also discuss the complex organization of the transduction molecules that underlie pheromone detection and signal transmission from vomeronasal neurons to the higher centers of the brain. Communication is a primary feature of living organisms, allowing the coordination of different behavioral paradigms among individuals. Communication has evolved through a variety of different strategies, and each species refined its own preferred communication medium. From a phylogenetic point of view, the most widespread and ancient way

A. Bigiani
Università di Modena, Dipartimento di Scienze Biomediche,
Modena, Italy

C. Mucignat-Caretta
Università di Padova, Dipartimento di Anatomia e Fisiologia Umana,
Padova, Italy

G. Montani · R. Tirindelli (✉)
Università di Parma, Dipartimento di Neuroscienze, Sezione di Fisiologia,
Via Volturno 39, 43100 Parma, Italy
e-mail: robertin@unipr.it · Tel.: +39-0521-903890 · Fax: +39-0521-903900

of communication is through chemical signals named pheromones: it occurs in all taxa, from prokaryotes to eukaryotes. The release of specific pheromones into the environment is a sensitive and definite way to send messages to other members of the same species. Therefore, the action of an organism can alter the behavior of another organism, thereby increasing the fitness of either or both. Albeit slow in transmission and not easily modulated, pheromones can travel around objects in the dark and over long distances. In addition, they are emitted when necessary and their biosynthesis is usually economic. In essence, they represent the most efficient tool to refine the pattern of social behaviors and reproductive strategies.

Detection of pheromones

In the animal kingdom, the wide use of pheromones for communication has raised the necessity of increasing the chance of detection of these intraspecific signals. Consequently, throughout evolution, many species have developed appropriate sensory systems that are almost exclusively devoted to the recognition and interpretation of the sexual/social information contained in pheromones.

Pheromones in mammals

The importance of social behavior for the survival of the species and of the individual has long been recognized. In the 1950s, the entomologists Karlson and Lüscher (1959) proposed the use of a new term to indicate substances, secreted by an individual, that release a specific reaction in another member of the same species. Since then, the term "pheromone" has been used to indicate almost all the molecules that are employed in communication, even if they do not fully accomplish the definition's requirements. In fact, some putative pheromones can be regarded as social odors, which allow, for example, kinship identification. These chemicals require suitable contextual information to exert an effect (Johnston 1998). In addition, when pheromones are used as messages among different species, for predation and defense purposes, these are often referred to as "allomones" (Brown et al. 1970).

Pheromones can be broadly grouped according to their functions so that, for example, trace pheromones encourage following a conspecific, while alarm pheromones alert other conspecifics about an external challenge, and can eventually evoke aggressive reactions.

A useful distinction is made between releaser and primer pheromones. Releaser pheromones elicit a short-latency behavioral response in the receiver, usually involving an interaction between two individuals, like in aggressive attacks or mating. Primer pheromones induce delayed responses that are commonly mediated through the neuroendocrine system. In this way, it is possible for an animal to modulate the reproductive status of another.

Tricky terminology has become popular among students in the pheromone field. It refers to the nature of the pheromone and aims to dichotomize chemical signals in "volatile" and "non-volatile" categories. While molecular mass and chemical features can be predictive of the "volatility" of a single molecule, the effective possibility for a molecule to reach the appropriate sensory organ cannot be readily prognosticated. Dusty particles or pollens can easily reach our nostrils in the air stream—as each allergic person can confirm—nevertheless their mass is usually larger than a single molecule, or even a protein. In addition, every chemical sensor in mammals is not directly exposed to the air stream; instead, a mucous layer, which must be crossed by molecules in order to be sensed, covers it. This adds a further level of intricacy to the slippery issue of volatility. Therefore, in our view, this misleading classification should be abandoned.

In mammals, pheromones have been implicated in a variety of effects in distantly related species. In marsupials, estrus can depend on pheromonal stimuli (Fadem 1987) and stern gland marking is popular among opossums, while insectivores, like moles and hedgehogs, use chemical signals to communicate among themselves. Both carnivores (dogs and cats) (Doty and Dunbar 1974; Goodwin et al. 1979; Verberne 1976) and ungulates (e.g., deer) mark their territory; in pigs, social hierarchy is modulated by pheromones (McGlone 1985) and even some primates (*Lemur catta*) engage in olfactory battles. The paradigmatic model for mammalian pheromone investigation involves laboratories species, among which rodents (hamster, mice and rats) are the most studied. In these species, several effects have been carefully described, and the mechanisms of their action dissected.

In hamster, the most well-known effect involves a female pheromone that induces a copulatory response in males. The pheromone was identified as the protein aphrodisin, which is secreted along with the vaginal fluid (Singer et al. 1987). Aphrodisin belongs to the superfamily of lipocalins that are secreted molecules able to accommodate and transport hydrophobic ligands in aqueous media. Lipocalins share a common motif (glycine-X-tryptophan) and have a similar structure, consisting of an eight-stranded β-barrel and an α-helix, with loops connecting the β-strands (Cavaggioni and Mucignat-Caretta 2000).

In rats, several pheromones are used in different contexts. Territory marking (Richards and Stevens 1974), alarm (Abel 1991), and maternal pheromones (Leon and Moltz 1973; Moltz and Lee 1981) have been described.

Mouse ecology heavily depends on pheromonal signals, which have been investigated for a long time. Confining the description to primer pheromones, it is known that grouped females modify or suppress their estrous cycle (Van Der Lee and Boot 1955), while male urine can restore and synchronize the estrus cycle of non-cycling females (Whitten 1956) or accelerate puberty onset in females (Vandenbergh 1969). In addition, the exposure of a recently mated female mouse to a male other than the stud prevents implantation of fertilized ova (Bruce 1960), implying that the stud or its odors are memorized at the moment of mating in order to be recognized later.

All these effects involve the release of chemicals with urine. Some low molecular weight molecules that possess pheromonal activity are typically found in female (Novotny et al. 1986) and in male urine (Schwende et al. 1986). Among the several androgen-dependent substances that have been identified in male urine, both 2-sec-butyl-4,5-dihydrothiazole and 3,4 dehydro-exo-brevicomin were reported to modulate intermale aggression and territory marking (Novotny et al. 1990; Novotny et al. 1985). Interestingly, the former molecule has optical activity, but only one stereoisomer is detectable in urine (Cavaggioni et al. 2003).

Male urine of rodents contains an unusually high quantity of protein, termed major urinary proteins (MUP) in the mouse and α-2 urinary proteins (α_{2u}) in the rat. They are both androgen-dependent, synthesized in the liver, filtered by the renal glomeruli, and finally excreted with urine (Cavaggioni and Mucignat-Caretta 2000). That being the case, MUP and α_{2u} are also expressed in such exocrine glands as mammary, parotid, sublingual, submaxillary, lachrymal, and nasal, and in modified sebaceous glands like preputial and perianal (Shahan et al. 1987). Both MUP and α_{2u} belong to the lipocalin superfamily and bind odorant molecules with affinity ranging from 10^{-4} to 10^{-6} M. Thus, they are likely to transport hydrophobic urinary pheromones in an aqueous medium (Bacchini et al. 1992; Bocskei et al. 1992). Several pheromonal effects have been attributed to MUP and α_{2u}. For example, these proteins, with their bound urinary pheromones, attract adult females while repelling males (Mucignat-Caretta et al. 1998). Other effects of MUP include the modification of light-avoidance behavior in both male and female mice (Mucignat-Caretta 2002; Mucignat-Caretta and Caretta 1999b), the aggressive reactions of males towards adult and newborn mice (Mucignat-Caretta and Caretta 1999a; Mucignat-Caretta et al. 2004), and the acceleration of puberty onset in female mice (Mucignat-Caretta et al. 1995).

This complex representation of pheromonal effects raises the question of how these chemicals are sensed by peripheral organs. The main olfactory system is principally devoted to the perception of odorants that can signal food and danger. However, in a limited number of cases, intraspecific pheromones are detected by olfactory chemosensory neurons whose stimulation elicits typical species-specific behavioral effects (Dorries et al. 1997; Hudson and Distel 1986). The majority of pheromones are believed to carry their information through a specific sensory organ and related transduction system, namely the vomeronasal organ of Jacobson (VNO).

Anatomy of the vomeronasal organ

When talking about chemical senses, we usually refer to two main sensory organs, the tongue for taste and the nasal mucosa for smell. The neurosensory epithelium of the nasal mucosa lines the upper part of the turbinates and septum. It consists of a pseudostratified epithelium, in which supporting cells span the whole epithelium width. At the base of the epithelium, globose basal cells constitute the precursor pool that gives rise to new olfactory sensory neurons throughout life. Receptor neurons have, in fact, a finite life span, lasting a few weeks, after which they degenerate. They are then substituted by new developing neurons that protrude their ciliated dendrites towards the nasal lumen, and ultimately develop an axon directed to the main olfactory bulb (Buck 1996; Cowan and Roskams 2002; Getchell 1986). The sensory axons contact the dendrites of mitral cells in the glomeruli. Olfactory information is then transmitted to a variety of structures, including the lateral amygdala and piriform cortex, which relay information to the entorhinal cortex and, via the mediodorsal thalamus, to the orbitofrontal cortex. Yet, in the nasal cavity there are at least four other chemical sensory systems, namely the trigeminal nerve terminals, the terminal nerve, the septal organ of Masera, and the VNO (Masera 1943; Wirsig-Wiechmann et al. 2002). The trigeminal system is mainly involved in the perception of irritating substances like ammonia, while the function of the terminal nerve is unclear. The organ of Masera consists of a batch of sensory neurons within the respiratory mucosa in the ventrocaudal portion of the respiratory cavities at the base of the nasal septum, and its function is still unknown. The VNO was first described in 1813 by the Danish anatomist Ludvig Jacobson (an English translation of the report is now available in Jacobson et al. 1998). He described the organ, which was named after him, in a variety of domesticated Mammals, as well as in some wild carnivores, ungulates, and marine mammals. The VNO is absent in fish, birds, and crocodiles, but it appears to have evolved in amphibians. It is well developed in snakes, where it has a role in the tongue-flick, a trailing behavior. Most mammals possess a VNO, with some exceptions: In dolphins, for example it is apparently absent, and it is vestigial in Old World monkeys, in apes, and in *Homo sapiens* (Doving and Trotier 1998).

Different reviews that describe in detail the morphological structure and function of the VNO are available (Cavaggioni et al. 1999; Doving and Trotier 1998; Halpern 1987; Halpern and Martinez-Marcos 2003; Keverne 1999, 2002). A brief description follows.

The mammalian VNO is a bony-encased structure that lies bilaterally at the base of the septum (Fig. 1). It opens anteriorly into the nasal cavities or into the mouth through the vomeronasal duct; hence, chemical stimuli can enter from both nose and mouth.

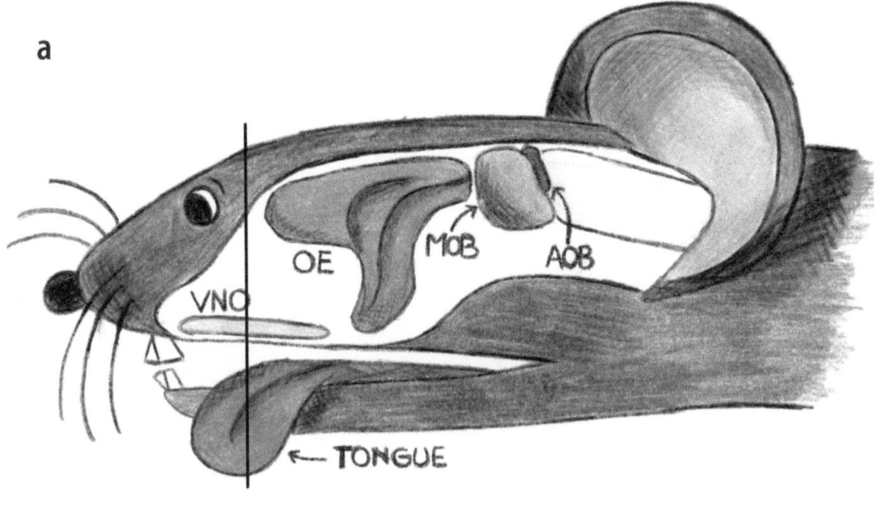

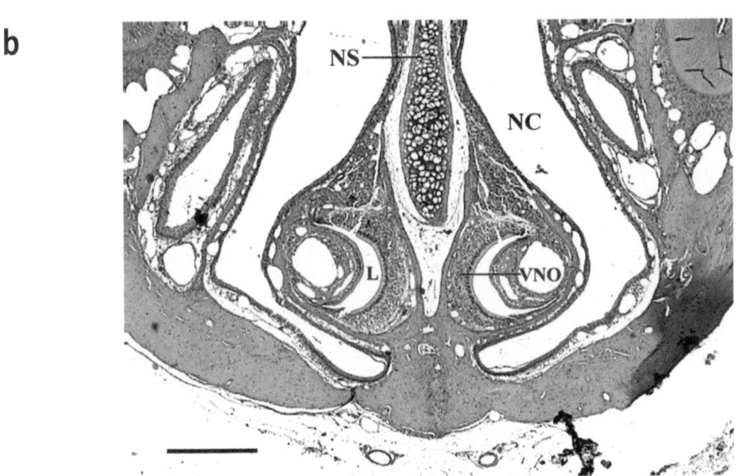

Fig. 1a, b Macro- and microanatomy of the olfactory organs in the mouse. **a** Drawing representing the position of the vomeronasal organ in the mouse head. *VNO* vomeronasal organ, *OE* main olfactory epithelium, *AOB* accessory olfactory bulb, *MOB* olfactory bulb. **b** Cross-section of a mouse head as indicated by Fig. 1a. *NC* nasal cavity, *NS* nasal septum, *VNO* vomeronasal organ. The *line* indicates the location of the sensory epithelium. The non-sensory epithelium lies on the opposite side, *L* is the lumen of the VNO. The lumen opens to the nasal cavity via the vomeronasal duct (not visible in this image). Due to the recessed location of the VNO, it is believed that both hydrophobic and hydrophilic pheromones can reach the microvilli of the sensory epithelium. *Scale bar*=200 μm

The organ is made of a spongy cavernous tissue that can swell and deflate, thus causing the stimuli to enter into the lumen of the organ. The VNO lumen, fully occupied by mucus, is crescent-shaped and covered by sensory epithelium in its medial portion and by non-sensory epithelium laterally (Fig. 1). The mucus is highly enriched in lipocalins that can possibly act as pheromone carriers (Miyawaki et al. 1994). Thus, stimulus access is not limited to airborne substances, but is available to all molecules, in particular to the largest ones (Wysocki

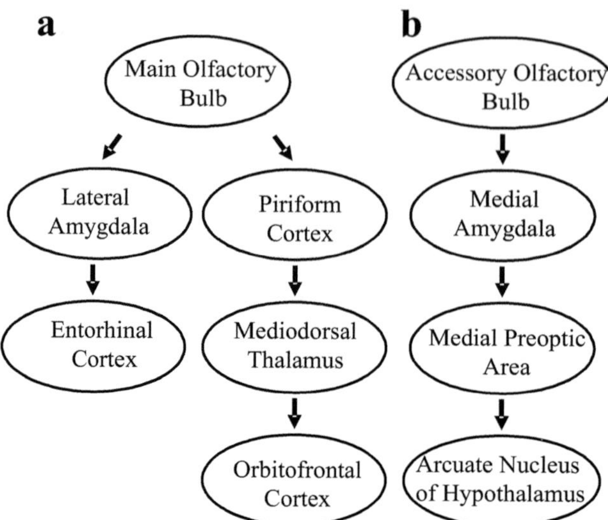

Fig. 2a, b Central areas interested by the vomeronasal and main olfactory systems. The diagram depicts the central pathways for the main (**a**) and accessory (**b**) olfactory systems. Only the principal relay stations are displayed. The connection between the different nuclei within and between the main and accessory projection areas is not illustrated. Note that the main olfactory system projects to telencephalic areas, in particular to high-order cortices, while the accessory olfactory system is mainly connected to the neuroendocrine hypothalamus

et al. 1980). The sensory epithelium hosts receptor neurons that project their apical dendrites to the lumen. On the opposite side of the sensory neuron, axons leave the mucosa and run parallel to the olfactory fibers in the first cranial nerve, crossing the lamina cribrosa of the ethmoid bone to enter the brain. The first synapse is in the accessory olfactory bulb (AOB), which lies dorsocaudally to the main olfactory bulb. Here, the vomeronasal axons reach the dendrites of the AOB mitral cells in structures called glomeruli. A sub-organization, however, exists in the vomeronasal system as neurons of the apical layer of the VNO send their axons to the anterior AOB, whereas basal neurons extend their axons to the posterior AOB. Mitral cells, with their dendritic pole, connect glomeruli of the same anterior or posterior zone whereas axons are routed to the medial amygdala, in which the information from both anterior and posterior zone converge (von Campenhausen and Mori 2000). From here, vomeronasal information travels to the hypothalamus, where input stimuli modulate the reproductive physiology (Fig. 2). At variance with the main olfactory bulb, which ultimately projects to prefrontal cortical areas, the tertiary projections of the VNO remain substantially subcortical, even if a cross-talk between the main and accessory olfactory modality is likely to take place already at the level of the amygdala (Halpern and Martinez-Marcos 2003). This can explain why some pheromonal effects are evident in inexperienced animals (for example, virgin animals in a mating contest), which have never faced a given situation (Westberry and Meredith 2003).

VNO-mediated behaviors

The importance of pheromone perception mediated by the VNO varies from one species to another, yet the consequence of VNO removal can ultimately affect the behavior of several animals (Wysocki and Lepri 1991). Nevertheless, it should be noted that, even within closely related groups, the relative extension of the VNO projection areas can vary greatly (Stephan et al. 1982).

In the opossum, the external chemical stimuli can be delivered to the VNO via nuzzling, a peculiar behavior that consists of frequent rubbing motions with the ventral aspect of the snout over the odor source (Poran et al. 1993). The stimuli perceived through the VNO can hence modulate the differential scent marking elicited by the conspecifics' odor (Shapiro et al. 1996).

It is believed that the "flehmen behavior" (a peculiar grimacing of the upper lips) of felines and ungulates has the function to deliver the chemical stimuli to the VNO (Doving and Trotier 1998).

Among sheep, the recognition of the lamb is fundamental for lactation and maternal care, and appears to be mediated by the VNO (Booth and Katz 2000).

The best-known effects were studied in rodents, in particular hamster, rats, and mice. A common finding in these species is that the vomeronasal system and its central projection are larger in males (Guillamon and Segovia 1997).

In hamsters, it has been shown that both the VNO and the vomeronasal central projection areas are activated in the male following stimulation with female hamster pheromones (Fernandez-Fewell and Meredith 1994; Kroner et al. 1996). This activation leads to a copulatory response. Moreover, the responses induced in the medial amygdala are different between sexes (Meredith and Westberry 2004). In the final central projection target area, which is the medial preoptic nucleus, the sensitivity to pheromonal stimulation and the responses to it are modulated by gonadal steroids (Swann 1997). Finally, the activation of the VNO in males is responsible for individual recognition (Johnston and Peng 2000).

In the rat, several aspects of social interactions are modulated by the VNO, which detects emitted pheromones according to age, gender, and social status. Maternal interactions are facilitated by chemical cues emitted by pups; the active substance has been identified as dodecyl-propionate secreted by preputial glands (Brouette-Lahlou et al. 1999). In addition, copulation is influenced by VNO stimulation (Saito and Moltz 1986), as male pheromones may induce ovulation in female rats (Johns et al. 1978).

In the mouse, the perception of pheromones through the VNO permits gender recognition (Wysocki et al. 1982) and leads to striking behavioral and hormonal outcomes.

The stimulation with pheromones induces differential activation in the anterior and posterior zone of the AOB (Dudley and Moss 1999), thus reflecting the heterogeneity of the afferents to these portions. The responses to pheromones of both the vomeronasal and central neurons are sexually dimorphic, also showing a specific strain selectivity (Halem et al. 1999; Holy et al. 2000; Luo et al. 2003; Luo and Katz 2004).

The most striking pheromonal effect that has been described in mice, namely the block of implantation after fertilization and the resultant failure of pregnancy, is indeed mediated by the VNO. The pheromonal stimuli from the VNO are decoded in the AOB, where they are memorized (Brennan et al. 1999; Okere and Kaba 2000).

Humans and pheromonal cues

The issue of the existence of human pheromones has long been posed (Comfort 1971). Reports show that some monkeys do indeed utilize pheromones in communication between sexes (Michael and Keverne 1970). It is also conceivable that olfaction has a role in the modulation of human neonatal behavior, due to its advanced stage of maturation, compared to other sensory systems at birth (Winberg and Porter 1998). Evidence points to the existence of male pheromones that modulate mood; one such pheromone was identified as androstenol (Benton 1982). More recent data show that male axillary extracts can modulate mood as well as the pulsatile secretion of luteinizing hormone (LH) (Preti et al. 2003). In addition, androstadienone, a putative male pheromone, activates the anterior portion of the inferior lateral prefrontal cortex and the posterior part of the superior temporal cortex. Both of these are cortical fields related to social cognition and attention (Gulyas et al. 2004).

Among the best-known effects attributed to human pheromones is the synchronization of menstrual cycle and the regulation of ovulation by armpit secretion of donor females (McClintock 1984; Stern and McClintock 1998). In addition, the exposure to human pheromones was shown to differentially activate the hypothalamus of males and females (Savic et al. 2001).

In all these studies, the question of which sensory organ detects pheromones is not addressed. Humans indeed possess a functional olfactory organ, whose sensing ability is often misunderstood. In contrast, the existence of a human VNO has been debated for a long time. The first reports date back to the middle of the nineteenth century (Bhatnagar and Smith 2003). Nowadays, a VNO is believed to be vestigial or absent in humans, being only present in fetuses and then disappearing before birth (Monti-Bloch et al. 1998). When identifiable in the adult, the human VNO consists of an invagination, termed vomeronasal pit, in the epithelium that lines the basal portion of the nostrils. The organ is covered by pseudostratified epithelium lying on a lamina propria and includes exocrine glands. Some cells of the epithelium exhibit microvilli and express neurofilaments (Jahnke and Merker 2000). There is neither evidence that microvillous cells send sensory axons to the olfactory bulb nor that a sizable AOB exists in humans. Nevertheless, it is plausible that some sort of pheromonal communication occurs via other characterized (main olfactory epithelium) or uncharacterized chemosensory organs. In this context, it should be remembered that cross-talk between the main and accessory olfactory system is present at the level of the amygdala, where a multimodal integrative chemosensory map can be formed (Fig. 2).

Vomeronasal receptors

The poor knowledge about the entity of pheromones and the not implicit correlation between the observed behavior in housed animals or in the wild represents the major barrier for the understanding of vomeronasal responses. However, objectively, for any response, stimulation of VNO chemosensory neurons by pheromones must occur when these molecules bind to specific peripheral receptors. In the main olfactory system, thousands of odorants are perceived on a psychophysical basis, and this panel of molecules largely exceeds the number of functional olfactory receptors (ORs) existing in the distinct mammalian genomes. This apparent discrepancy is solved as each receptor type binds multiple chemically related odors, and a single odor, in turn, stimulates more than one receptor type (Malnic et al. 1999). Translated into music, one can imagine that at the molecular level each odor is identified by an interval where the played notes are the distinct stimulated receptors and the sound produces the quality of the odor itself. Consonant intervals, say the perfect fifth, represent pleasant odors, like "violetta di Parma" or jasmine, whereas stinky odors are originated by dissonant intervals.

In the pheromonal system, the conscious discrimination of a pheromone is probably unnecessary, but crucial is the behavior that it elicits. Consequently, properties of pheromone receptors must be different from those observed in the main olfactory system. Unfortunately, our knowledge in this field is not yet complete, although a burden of data about the entities and distribution of vomeronasal receptors has been accumulating in the last 10 years (Rodriguez 2004).

Discovery

A milestone in vomeronasal studies was the description of the radial expression pattern of two G protein α-subunits, $G\alpha_o$ and $G\alpha_{i2}$ (Berghard et al. 1996; Jia and Halpern 1996). Their distribution was found to be complementary in that $G\alpha_{i2}$ is abundantly expressed in apical neurons and their associated axons, and $G\alpha_o$ exclusively in the basal ones (Fig. 3). No expression overlap between these transduction molecules was observed in the VNO. Thanks to the high expression level of $G\alpha_{i2}$ and $G\alpha_o$ in the axons of the chemosensory neurons, it was possible to follow the projections of these two populations of neurons and find that these synapsed into two distinct areas of the accessory olfactory bulb, namely the anterior and posterior AOB (Jia and Halpern 1996). The conclusion of this seminal study led to the identification of two parallel subsystems for the transduction of pheromonal stimuli. Both $G\alpha_o$ and $G\alpha_{i2}$ are absent in mature olfactory neurons that, as the alternative, express the G protein α-subunit $G\alpha_{olf}$ (Jones and Reed 1989). This molecular divergence reinforces the notion that an early phylogenetic separation occurred between the olfactory and vomeronasal system (Bertmar 1981). While the molecular and functional divergence of the vomeronasal system had aroused the interest of many neuroscientists, difficulties soon arose for the search of

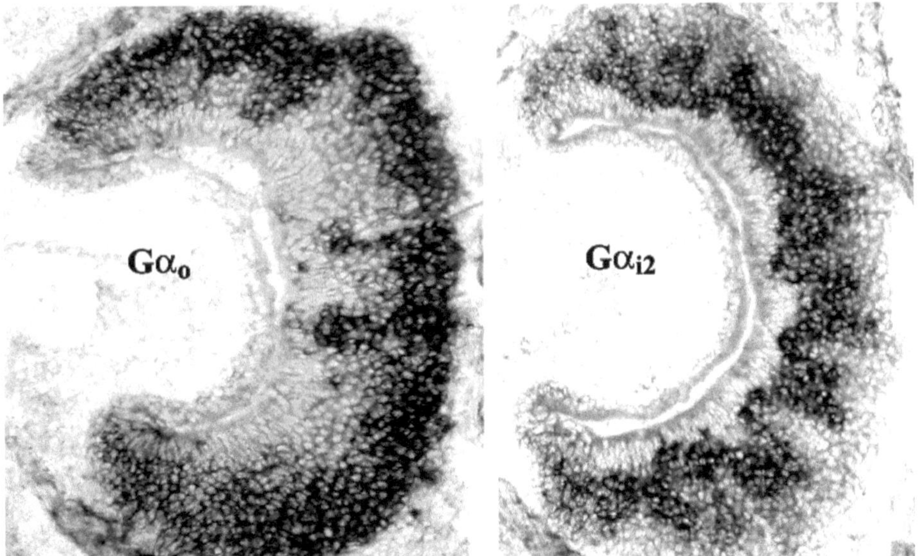

Fig. 3 The specific pattern of expression of $G\alpha_{i2}$ and $G\alpha_o$ in the VNO. The starting point of the molecular studies on the vomeronasal organ stems from the observation that two neuronal populations of the VNO express different G protein α-subunits. $G\alpha_o$-expressing neurons are located in the basal VNO and project to the posterior AOB. $G\alpha_{i2}$-expressing neurons can be identified more apically and project to the anterior AOB. Thus, the identification of different subsystems for the detection of pheromonal signaling suggests that at least two types of receptors must exist in the VNO. Putative pheromone receptors, V1Rs, co-express with $G\alpha_{i2}$ in apical neurons, whereas V2Rs do so with $G\alpha_o$ in the basal ones

other transduction molecules. In fact, attempts to identify putative pheromone receptors on the homology basis with the ORs brutally hit a rubber wall. ORs were discovered starting from the assumption that a weak homology among all G protein-coupled receptors (GPCRs) exists. Highly degenerated oligonucleotides were designed and successfully probed in polymerase chain reactions (PCR) starting from RNA of the whole olfactory epithelium (Buck and Axel 1991). This major discovery shed light on the largest receptor family ever known in a mammalian genome. Approximately 1,000 members were later identified in rodents but also in other species and in humans too, where the number of pseudogenes is approximately 50% of the total (Glusman et al. 2001; Niimura and Nei 2003; Young et al. 2003; Zhang and Firestein 2002).

A different strategy was then adopted by Dulac and Axel in an attempt to identify vomeronasal receptors (Dulac and Axel 1995). This was the result of elegant experiments based on the following simple assumption: Although the vomeronasal receptors were probably too divergent from the olfactory ones, similitude would persist between the two sensory systems. For example, in the main olfactory epithelium, an individual chemosensory neuron expresses only a single receptor type at high level (Vassar et al. 1994). If this was also true for the vomeronasal neurons, the mRNA of two randomly isolated neurons should share the same complexity and only differ for the presence of two distinct receptor sequences. cDNA libraries from single neurons were constructed and clones were cross-hybridized with cDNA of other individual neurons. Non-hybridizing clones were isolated and sequenced. From an extensive screening, a novel receptor family, named VNs, was identified and shown to be exclusively expressed in the vomeronasal neurons (Dulac and Axel 1995; Pantages and Dulac 2000). These receptors were later named, for simplicity, V1Rs (Tirindelli et al. 1998). Further studies were addressed to determine the extension of this family that was ultimately found to group more than 100 functional genes (Lane et al. 2002; Rodriguez and Mombaerts 2002). It also turned out that V1R expression is restricted to the apical neurons of the VNO that also express $G\alpha_{i2}$ (Dulac and Axel 1995; Pantages and Dulac 2000). Since none of these novel receptors extended their expression in the basal layer of chemosensory neurons, the same strategy was applied to the search for receptors that could be present there. This time, however, single cell libraries were made, but exclusively from selected vomeronasal neurons expressing $G\alpha_o$. The procedure again proved successful and a second family of receptors, named V2Rs, was brought to light (Herrada and Dulac 1997; Matsunami and Buck 1997). However, similar results were obtained with a more traditional approach to the problem (Ryba and Tirindelli 1997).

In conclusion, 3 years after the description of two distinct signal transduction pathways in the VNO, two different families of GPCRs were revealed (Fig. 4).

Structure

Unrelated by amino acid sequence, the two families of vomeronasal receptors are also structurally very different (Tirindelli et al. 1998). V1Rs are topologically related to group I of the GPCRs. Group I also includes other chemosensory receptors like olfactory and taste receptors for bitter substances (Adler et al. 2000; Buck and Axel 1991). V1Rs represent a relatively large family accounting for approximately 100 functional members in the mouse. They are grouped in 12 subfamilies according to sequence homology (Del Punta et al. 2000; Rodriguez et al. 2002). Interestingly, even in rodents that make a large use of pheromones for intraspecific communication, a high number of pseudogenes have been described (Ro-

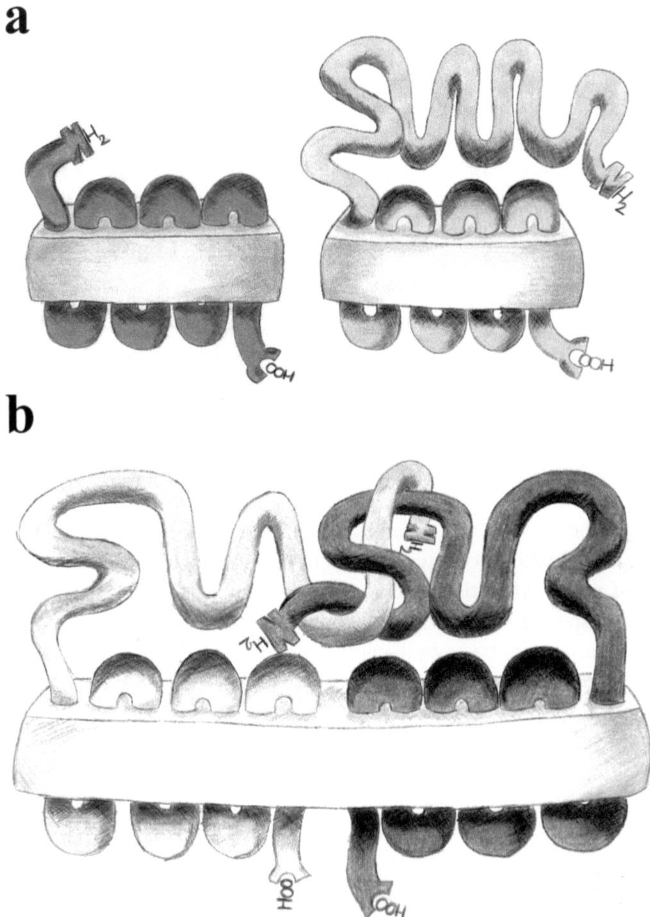

Fig. 4a, b Schematic organization of pheromone receptors. **a** V1Rs are related by structure to ORs as they both present short intracellular and extracellular regions. The binding site of V1Rs is believed to be positioned in a hydrophobic pocket of the transmembrane region, probably spanning between helices II and VII. V2Rs compared to V1Rs have an additional long extracellular region of approximately 70 kDa that is believed to represent the ligand-binding domain of the receptor. **b** According to the crystal structure of the related extracellular region of mGluR1α, V2Rs are likely to be folded as hetero ($V2R_x$–$V2R_2$) or homo ($V2R_2$) dimers on the surface of the plasma membrane

driguez et al. 2002), and like for the ORs, pseudogenes are probably highly expressed in the vomeronasal neurons, at least at the mRNA level (Serizawa et al. 2003).

Where do V1Rs differ from each other? It is plausible to predict that the putative pheromone binding sites show the highest variability among these receptors as they sense pheromones, which, in many cases, display remarkable chemical differences. As a comparison, a quick description of the structural features of the ORs may be helpful for the reader.

ORs have been built to detect thousands of odorants and, probably, to link to only one species of G protein, the olfactory-specific $G\alpha_s$, $G\alpha_{olf}$ (Jones and Reed 1989). Although many ORs are still orphan of their agonists, a region lying between the transmembrane (TM)2 and TM7 has been identified as the ligand-binding domain of these receptors. First,

ORs are more divergent in these regions than in other domains; second, the sequence between TM2 and TM7 of odorant receptors, when replaced with that of another GPCR, imparts odor specificity to the resulting chimeric protein (Krautwurst et al. 1998); third, All ORs that sense related molecules show a high degree of homology in the TM2–7 region (Buck and Axel 1991); fourth, a point mutation in the amino acid sequence of this region changes the affinity of ORs for ligands (Krautwurst et al. 1998). A quick survey on V1Rs suggests that similar features should characterize these receptors. As for most odorants, the known pheromones (at least) are hydrophobic molecules, and, consequently, a binding site folded in the transmembrane region of these receptors is somewhat envisaged (Fig. 4).

Unfortunately, the major problem with V1Rs is their reluctance to be expressed by heterologous systems and this barrier cannot apparently be overcome as it can for ORs. However, clues on the molecular properties of V1Rs were obtained from a physiological approach to the problem. Calcium imaging experiments on living VNO sections concluded that any specific apical neuron (presumably a V1R-expressing neuron) is stimulated by a single pheromonal molecule (Leinders-Zufall et al. 2000). This contrasts with the flexibility of the olfactory neurons that are stimulated by many molecules.

A specific pheromone for a V1R (V1R2b) has been identified recently by Boschat and colleagues (Boschat et al. 2002). The way that these authors came to this conclusion was via the co-expression of V1R2b along with green fluorescent protein (GFP) in VNO neurons of knock-in lines of mice. After tissue dissociation, individual fluorescent neurons were isolated and tested for the activation with various pheromones by a calcium imaging approach. The verdict was that heptanone, a well-known pheromone that contributes to induce aggression in males and puberty in females (Novotny et al. 1986, 1999), was the only molecule able to elicit a physiological response in the GFP-expressing neurons. In essence, although arduous to identify, there might be critical structural differences between ORs and V1Rs. ORs are likely to possess a flexible binding pocket that can accommodate several ligands, whereas rigidity should be a feature of V1Rs. Unfortunately, all predictions that can be formulated on the structural basis of V1R receptors cannot find confirmation until a system for expressing these receptors is developed.

The distribution pattern and structure of V2Rs perhaps deserve more specific attention. V2Rs, in fact, belong to the group 3 GPCRs, a class of molecules that includes several receptors like the extensively studied metabotropic glutamate receptors (mGluRs) (Pin and Duvoisin 1995), γ-aminobutyric acid receptor (GABA)$_B$s (Kaupmann et al. 1997), and calcium sensing receptor (CaSR) (Brown et al. 1993). However, chemosensory receptors like taste receptors for sweet molecules and amino acids (T1Rs) (Hoon et al. 1999; Matsunami et al. 2000; Max et al. 2001; Montmayeur et al. 2001; Nelson et al. 2001, 2002), fish ORs (GBRs)(Cao et al. 1998; Speca et al. 1999) are also members of group 3.

The V2R family accounts for approximately 100 members, but the real extension of this family is difficult to decipher as the rodent genome contains several pseudogenes that, similar to V1Rs and ORs, are efficiently transcribed in individual neurons. The way in which these receptors were discovered, at least by two of the three research groups that succeeded in this effort (Herrada and Dulac 1997; Matsunami and Buck 1997), left no doubt that V2Rs, like ORs and V1Rs, are individually expressed in single chemosensory neurons. However, the identification of V2Rs by a third research group (Ryba and Tirindelli 1997), simply making use of PCR and degenerate oligonucleotide probes, was somewhat instructive for the knowledge of the extension of this receptor family. One of the receptors that they isolated, namely V2R$_2$, was found to be expressed in all basal neurons and to co-express in the same neuron as other V2Rs (Martini et al. 2001). Sequence comparisons of V2Rs also demonstrate that V2R$_2$ is a divergent member of this family and places it closer to fish ORs than to other rodent V2Rs (Martini et al. 2001). Thus, the peculiar expression pattern of V2Rs in the vomeronasal system parallels that of taste cells. Here, three receptors (T1R1, T1R2, T1R3) have been identified for sensing sweet molecules and amino acids. The co-expression of T1R3 with one of the other two confers the specificity to the taste cell to respond to amino acids (T1R1) or sweet molecules (T1R2) (Hoon et al. 1999; Nelson et al. 2001, 2002; Zhang et al. 2003; Zhao et al. 2003). Similarly, the co-expression of $V2R_2$ with another V2R might be necessary for the binding of pheromones.

The question of interest is how these receptors interact to accomplish this task. Before speculating, it is important to describe how group 3 GPCRs are presented onto the surface of the plasma membrane. The main feature of receptors of this group is a large (70 kDa) extracellular region that is responsible for the binding with their ligand (Pin and Duvoisin 1995) (Fig. 4), binding that also occurs when the extracellular domain is expressed without the transmembrane region (Okamoto et al. 1998). Taking advantage of the solubility of this ligand-binding region, several biochemical studies led to some important conclusions. First, these receptors are fully functional only in a dimeric form (Okamoto et al. 1998; Sato et al. 2003; White et al. 1998); second, dimerization occurs via hydrophobic domains in the extracellular region and it is stabilized by cysteine linkages (Tsuji et al. 2000); third, the extracellular regions are folded in such a way that the complex is shaped according to a "Venus flytrap" model that allows ligand binding (Jingami et al. 2003; Kunishima et al. 2000). Finally, but not less intriguing is the fact that all these receptors show the feature of binding amino acids (Conigrave et al. 2000). Also CaSR, which is described as a sensor of calcium in the body fluids, appears to be modulated by amino acids. The fish OR 5.24 binds basic amino acids (Speca et al. 1999); PIR1+T1R3 and T1R2+T1R3 bind L or D amino acids (Nelson et al. 2002; Zhao et al. 2003).

Structural analysis of group 3 GPCRs has identified potential ligand docking residues that are conserved in many receptors of this family. The key residues that interact with α-amino and α-carboxyl groups are all conserved in the amino acid binding receptors CaSR, R5.24, and T1R1+T1R3 (Kuang et al. 2003; Kunishima et al. 2000). Curiously, T1R2 that senses simple sugars lacks two of these residues.

In the VNO neurons, it is also likely that V2Rs dimerize to be fully functional (Martini et al. 2001). Nevertheless, whether they are associated on the plasma membrane as homodimers or heterodimers has not yet been proved. It is noteworthy, however, that $V2R_2$ but not other V2Rs conserves potential residues for docking amino acids (Silvotti et al. 2005), thus suggesting that amino acids or related molecules may bind to this receptor. This would imply a role for these molecules in pheromonal communication.

Very recently, two research groups reported the expression of major histocompatibility complex (MHC) class Ib molecules in V2R-expressing chemosensory neurons of the VNO (Ishii et al. 2003; Loconto et al. 2003) . Genes encoding MHC Ib molecules are clustered at the telomeric end of the mouse H2 locus and fall in two subfamilies of respectively eight and three members. As a consequence, V2Rs are co-expressed on the plasma membrane of basal neurons with a preferential combination of MHC molecules along with β2-microglobulin (β2m), thus forming a large multimolecular complex (Loconto et al. 2003). The interesting question is whether MHC molecules are directly involved in pheromonal communication (Hegde 2003). Are they essential to expose specific antigens to the plasma membrane? What is ultimately the function of the large combinatorial multimeric complex (MHC–β2m–$V2R_x$–$V2R_2$) that is formed on the plasma membrane of the vomeronasal neurons? Several studies have suggested that MHC genes impart the olfactory identity of each individual of the same species (Yamazaki et al. 1976). Additionally, MHC-dependent discrimination of mice of the same strain is a pre-determined behavior and it is not due to experience as confirmed by mice grown from embryo transfer in unrelated foster mothers (Isles et al. 2001). It is conceivable, for example, that pheromonal signaling in the basal neurons of the VNO is triggered by the mutual combination of self-molecules and pheromones acting on different points of the MHC–V2Rs complex. What the nature of self molecules is in pheromonal communication remains a very elusive issue (Pearse-Pratt et al. 1998; Singer et al. 1997). However, it is interesting that MUP show a combinatorial diversity of expression in urine at least as great as MHC Ib in the VNO (Hurst et al. 2001). In this context, it would be instructive, for example, to determine specific differences in the MHC Ib sequences of rat and mouse where identity among the V2R sequences of these species is unexpectedly high.

In conclusion, all these studies emphasize the growing complexity of the pheromone signaling mechanism in mammals and perhaps suggest that some of the concepts about pheromone signal transduction necessitate some revision.

Genomic organization

Information from the study of the genomic organization of the vomeronasal receptors concurs to define the biological role of the vomeronasal system and provide hints on the evolutionary processes undergone by many species. How does, for example, the repertoire of vomeronasal receptors adapt to fulfill species-specific requirements? The mouse genome has been revealed (Waterston et al. 2002) and DNA sequences are now available from public and private databases. Additionally, the absence of introns in the coding sequence of V1Rs and their clustering on chromosomes makes them an ideal subject for such studies (Del Punta et al. 2000; Lane et al. 2002; Rodriguez et al. 2002). At first glance, the extension of this family appears surprising: approximately 300 receptors grouped in twelve subfamilies, although the effective number of functional receptors is at least halved, given the high proportion of pseudogenes or incomplete receptor sequences. The V1R subfamilies are phylogenetically divergent, revealing an amino acid identity of only 15%–40%. A greater identity, approximately 40%–70%, however, appears when comparison is made between members of each subfamily. Some V1R loci in the mouse genome clearly reveal a history of gene duplication events that lead to significant changes in the repertoire of these receptors.

The accurate analysis performed by Lane et al. at three V1R loci located in two different chromosomes suggests that gene-block duplications occurred recently during mouse evolution and allowed the expansion of the V1R family (Lane et al. 2002). By analyzing the divergences between the duplicated blocks and the density of repetitive elements at individual loci, it is evident that duplication occurred, in bursts, not later than 20–30 million years ago, a time frame overlapping with that of the segregation between mouse and rat. It is plausible that changes that followed gene duplication constituted the existing reproductive barrier between these two related species.

However, the interesting information gained from these studies remains elusive since the knowledge about the nature of mammalian pheromones is so limited. Indeed, the few identified pheromones appear to be chemically very different, thus coinciding with the divergence between the subfamilies of V1Rs. If we assign to vomeronasal receptors the exquisite function to perceive sex-related molecules, the main question remains open as to whether the oversized repertoire of V1Rs reflects the evolutionary pressure exerted on individual receptors or on the mechanisms that underlay the expansion of these receptors.

The vomeronasal transduction mechanism relies on the ability of each sensory neuron to transcribe only a single receptor gene from one of the two alleles (allelic exclusion) (Belluscio et al. 1999; Chess et al. 1994; Del Punta et al. 2002b; Rodriguez et al. 1999). In addition, receptor expression is spatially regulated such that a given V1R or V2R is specifically expressed in one of the two specific zones of the VNO ($G\alpha_{i2}$ or $G\alpha_o$). Are there common regulatory sequences in the vomeronasal receptor genes that contain this information? Studies limited to 15 V1R sequences residing at a specific locus of chromosome 6 identified two conserved regions at respectively 1 kb and 5 kb upstream of the transcriptional start site (Lane et al. 2002). The homology of these sequences contrasts with the divergence in the coding regions of V1R in the same locus. More peculiar is the length of sequence conservation of these putative regulatory regions that suggests an unusual mechanism for controlling gene expression. For example, it is possible that these are candidate regions to control gene duplication or co-regulate gene expression in the same locus (locus control regions). In this context, the study of the OR genes should be helpful to define elements that control the zonal distribution, neuronal expression, and allelic exclusion of V1Rs. In fact, all these features are shared by ORs and V1Rs. However and very surprisingly, in the OR family, receptors do not share common motifs flanking all genes, although correct zonal expression of these receptors is observed with transgenes containing as little as 2.0 kb upstream of the transcriptional start site (Vassalli et al. 2002). Overall, this either suggests that V1Rs evolve independently from ORs or that common but novel mechanisms are operative in regulating gene expression.

Studies of the evolution of V2Rs are complicated by the presence of at least five intronic sequences that greatly increase the length (approximately 20 kb) and complexity of individual genes. It is undoubtedly hard, at present, to exactly define the size of the family, given also the surprisingly high density of pseudogenes (70% of the total), truncated receptor sequences, or interspersed exons in the rodent genome (Matsunami and Buck 1997). Although very little is known about the regulation of these genes, phylogenetic studies limited to the coding sequence of a small number of V2Rs in rat and mouse have remarked on the presence of amino acid sequences in the extracellular binding region of these receptors where a positive (Darwinian) selection indeed occurred (Emes et al. 2004). It is tempting to speculate that the rapid variation at these sites reflects the amplification and diversification of domains that are involved in the binding with structurally different pheromones. Ligands for V2Rs have not yet been identified; however, it is plausible that, like for all agonists of class 3 GPCRs, they are water soluble rather than hydrophobic molecules as predicted and partially demonstrated for V1Rs (Boschat et al. 2002; Leinders-Zufall et al. 2000). Contextually, lines of evidence strongly suggest that MUPs and α_{2u} arose via separate gene expansion (Clark et al. 1985). Moreover, the finding that conservation in α_{2u} is higher in the intronic rather than in the exonic sequences once again strengthens the hypothesis of a rapid coding sequence diversification (Emes et al. 2004). A phylogenetic analysis of the MUP sequences reveals that positive selected sites map to the surface of the protein rather than in the pheromone-binding pocket (Bocskei et al. 1992; Emes et al. 2004). This observation fits well with biochemical evidence that indicates the low specificity of members of the MUP family for hydrophobic ligands (Cavaggioni et al. 1990; Darwish Marie et al. 2001). A hazardous speculation would interpret these data as to whether the presence of positively selected superficial sites in MUP reflects those identified in the putative ligand-binding region of V2Rs, thus leading to assume that V2Rs interact with urinary protein pheromones.

What about human putative pheromone receptors? As described above, pheromonal effects involve humans too, although the presence of a functional VNO in this species is, at minimum, doubtful. However, in our genome, we have significant traces that suggest that our ancestors made a large use of pheromones to govern their social life. In fact, a total of 200 human V1R-like sequences (V1LRs) have been described in all chromosomes (except chromosome 20) (Giorgi et al. 2000; Rodriguez and Mombaerts 2002). Surprisingly, among these, five possess an intact open reading frame. Like in rodents, human receptors fall into distinct clades on phylogenetic trees, and, very interestingly, conserve 14 amino acid residues and a glycosylation site that have been described to be a signature of the V1R family. Overall, however, they display a weak homology when compared to mouse V1Rs. Together, these observations suggest a biological function of V1LRs in humans. Where is then the peripheral repository of pheromone receptors? Indeed, it is not in the vestigial VNO. One of the human receptors has been reported to be expressed in the olfactory mucosa (and other tissues), suggesting that the main olfactory system perceives conscious and unconscious chemical stimuli (Rodriguez et al. 2000). However, these data need to be handled with caution. First, mRNA expression is not a direct indication of protein expression in a given tissue, especially in the olfactory tissues where the mRNA of pseudogenes is abundantly expressed. Second, in the mouse, some common odorants are able to stimulate neuronal populations of the VNO through as-yet-unidentified receptors (Trinh and Storm 2003). It cannot be excluded that V1LRs, and possibly some rodent V1Rs, only represent a subpopulation of odorant receptors.

Molecular genetics

One of the fascinating approaches to understand the processing of information in the olfactory system was developed by Axel's group in 1996 (Mombaerts et al. 1996). This was based on the assumption that if each neuron of the main olfactory system expresses a single receptor type then it would become possible to follow the route of its axon to the olfactory bulb. At that time the crucial question was how the wiring network that connects the olfactory neurons to the olfactory bulb was conceived to allow us to perceive odorants. "Gene targeting" was the magic word and the successful approach that linked traditional neuroanatomy to modern molecular genetics. This task initially consisted in coupling the expression of a tracing molecule with that of a given receptor. The effort was materialized by creating knock-in mice in which the sequence encoding the microtubule-associated protein, Tau, was fused to that of the β-galactosidase gene (*LacZ*) and the product was inserted into the OR P2 locus by homologous recombination. Targeting vectors were designed that would result in a bicistronic RNA, thus assuring the expression of a functional receptor along with the tau–lacZ fusion protein. Thus, mice expressing an OR also express lacZ in the axons, making the visualization of the pattern of projections in the brain possible. An enormous amount of information has been squeezed out from this apparently simple technique. First, it was demonstrated that, although neurons expressing a given OR are scattered in one of the four zones of the olfactory epithelium, their axons converged with precision upon to a small number of glomeruli of the OB (normally two for each bulb; Mombaerts 1996; Mombaerts et al. 1996). In essence, given that the position of individual glomeruli are topographically defined (with some exceptions, see Zou et al. 2004), the bulb provides a spatial map in which each activated glomerulus corresponds to the stimulation of a specific receptor in the olfactory mucosa. Therefore, the quality of an odor is distinguished by the different pattern of brain activity (Malnic et al. 1999). Second, the establishment of a fixed topographic map is secondary to the nature of the specific receptor that is expressed in a given olfactory neuron. The replacement of the sole OR coding sequence with that of another receptor alters the axonal convergence and axons are rerouted to a couple of different glomeruli (Feinstein and Mombaerts 2004). The deletion of a receptor coding sequence in a specific locus or its partial silencing (low level of expression) has also dramatic effects on its axonal convergence, and fibers of the mutant neurons are broadly dispersed within regions of the outer nerve layer of the bulb without a detectable convergence (Feinstein et al. 2004; Wang et al. 1998). Third, each olfactory neuron stochastically chooses one of the two alleles to express a given receptor, and a functional receptor expressed in a given neuron contributes to the exclusion of another receptor being expressed in the same cell (Serizawa et al. 2000, 2003). All these results show that, in the main olfactory system, projections may segregate according to the receptor that they express, suggesting that the quality of odorants is encoded by a spatial map of neural activation.

Are all these mechanisms shared by the vomeronasal system? Does the vomeronasal system require such a wiring diagram to discriminate the quality of different pheromones? Given the nature of the ORs, the spatial map of glomerular activation for a given odorant is flexible, depending also upon the concentration of the odorant itself. If the vomeronasal system developed to trigger innate behaviors, conceptually, we would expect a more stereotyped map in which a pheromone always activates the same pattern of glomeruli. However, two aspects suggest that this minimalist model cannot be entirely taken into account. First, pheromones elicit their effect as a blend rather than as individual molecules (Hildebrand 1995). Second, there are not exclusively male or female pheromones but pheromones produced at different concentration in either sex (Holy et al. 2000). In essence, the pheromonal system, similarly to sounds in the inner ear, decodes complex information that is dependent upon the relative concentration of the molecules in it contained.

But let us describe what the most recent experimental data suggest. In the VNO, the receptor repertoire is much smaller than in the main olfactory epithelium (MOE); however, both V1Rs and V2Rs (with the exception of $V2R_2$) are individually expressed in chemosensory neurons, allowing the use of the gene targeting technique to identify the map of their projections (Belluscio et al. 1999; Del Punta et al. 2002b; Rodriguez et al. 1999). When the tau–lacZ (or tau–GFP) was inserted downstream of an individual V1R or V2R gene, the visualization of the axonal projections of neurons expressing that gene suggests that several glomeruli in subregions of the anterior or posterior AOB received inputs. Although glomeruli in the AOB are too loosely delimited to provide an accurate count, an estimate of 15–30 glomeruli for neurons expressing a given V1R and 5–7 for neurons expressing a given V2R probably represents a cautious evaluation. Like in the main olfactory bulb, each glomerulus appears uniquely innervated by axons expressing the same V1R or V2R receptor. Thus, it appears evident that the vomeronasal subsystems ($G\alpha_o$ and $G\alpha_{i2}$ dependent) use a general scheme to transduce the pheromonal stimulus to the brain.

Although this review is principally focused on the peripheral events of pheromonal reception, we cannot ignore the contribution of the molecular genetics to the study of the central connections in the vomeronasal system. Intriguingly, when gene targeting was linked to ordinary neural tracing, it turned out that dendrites of a mitral cell preferentially connect glomeruli expressing the same receptor (Del Punta et al. 2002b). In essence, a given pheromone exclusively stimulates a distinct population of neurons expressing the same receptor. Action potentials propagating along axons of the stimulated neurons converge to 5–30 glomeruli. Stimuli then proceed to secondary neurons, the mitral cells, in such a way that in a single mitral cell are conveyed inputs from several activated glomeruli. Overall, in the vomeronasal system, we are in the presence of a gradual convergence of the peripheral inputs. It is possible, for example, that axonal projections of neurons expressing a given receptor towards several glomeruli (rather than two as in the MOE) provide an indication of the concentration of a pheromone in a given blend. It would be interesting to ascertain if the fibers that converge to a group of glomeruli reflect the activation of receptors located at different rostrocaudal positions of the VNO. In fact, taking into account the peculiar anatomy of the VNO, it is conceivable that an antero-posterior concentration gradient of stimulatory molecules develops in the VNO when an animal is challenged with a pheromonal cue. Chemosensory neurons expressing a given receptor in the anterior VNO, which is closer to the vomeronasal duct, would be more stimulated than the distal ones and would send action potentials to a restricted subset of glomeruli. Consequently, mitral cells would then be able to spatially or temporally integrate and transmit signals to the higher centers from differentially activated glomeruli. Although there is no evidence for such a model, it is tempting to speculate that this accords with the lack of a consistent sex-specific expression of receptors in the VNO (Del Punta et al. 2002b; Matsunami and Buck 1997; Ryba and Tirindelli 1997) and with the observation that the concentration ratio of pheromones differs as much as a thousand fold in either sex (Holy et al. 2000).

Alternatively, as suggested by Del Punta et al. (2002b), the formation of multiple adjacent glomeruli receiving inputs from the same receptor provides a mechanism for enhancing the noise-to-signal ratio via lateral inhibition. This has also been observed in the main olfactory system, where the best-characterized inhibitory circuit in the MOB involves lateral inhibition of mitral cells at mitral–granule cell synapses in the external plexiform layer (Mori et al. 1999). However, intraglomerular circuits also form a center–surround organization in that the activation of a glomerulus causes widespread inhibition of mitral cells in neighboring glomeruli (Aungst et al. 2003).

Outputs from AOB and the activation of neuroendocrine pathways also represent an unsolved question. In the main olfactory system, a stereotyped olfactory map is represented in different cortical areas. Inputs from a stimulated OR are routed onto overlapping clusters of neurons at limited sites (Zou et al. 2001). Such an arrangement would favor an integration of the olfactory signals that may be important for odorant perception. Moreover, the defasciculation of the olfactory bulb outputs to different cortical areas may reflect the ability of the central nervous system to simultaneously activate different functions from a distinct olfactory stimulus.

Is it conceivable that this complex organization may attain the means of pheromonal perception? Indeed, the dichotomy that characterizes the peripheral organization of the vomeronasal system of rodents seems to persist in the higher centers of the brain (Simerly 1990). It would be interesting, in this context, to trace afferents from neurons expressing a given receptor and see if they ultimately project to large or specific subcortical areas. The conscious perception of a pheromone is thought to be unnecessary, and behaviors elicited by pheromones are presumably only a few. It is therefore expected that relatively wide areas receive inputs from different receptors. However, the main question remains open as to whether the two classes of receptors send stimuli to different regions of the brain with distinct functions or perhaps the existence of two pathways allows integrative responses to the stimulation with different pheromones.

Molecular genetics supplies us with additional information about the role of the VNO in pheromonal communication. For example, knocking out a large genomic region of the mouse, containing two entire subfamilies of V1Rs (16 functional members), results in a lack of electrical response of the VNO to 6-hydroxy-6-methyl-3-heptanone, n-pentyl acetate, and isobutylamine and precludes the mutated vomeronasal neurons from converging onto glomeruli of the AOB (Del Punta et al. 2002a). Additionally, mutant mice display specific dysfunctions such as reduced mounting behavior and maternal aggression. The latter effect is also confirmed in mice where the gene for the cation channel TRPC2 that triggers all electrical responses of the VNO, is deleted (Leypold et al. 2002; Stowers et al. 2002). However, this sort of molecular ablation of the VNO fails to show deficits in the reproductive responses as observed in V1R knock-out mice. It is possible that the total inactivation of the VNO causes other sensory pathways (olfactory perhaps) to act as default mechanisms to regulate sexual behaviors. On the other hand, the deletion of a small subset of receptors as performed by Mombaerts and colleagues (Del Punta et al. 2002a) may impair a predetermined map of pheromone activation in the brain, resulting in the appearance of distinct phenotypes.

Microphysiology of pheromone reception

Vomeronasal neurons (VNO neurons) are the peripheral detectors of pheromones. Their dendrite reaches the lumen of the VNO where pheromone-containing fluids are conveyed from the nasal cavity through an active vascular pumping mechanism controlled by the autonomic nervous system (Meredith 1998). As reported above, pheromones are thought to interact with chemical receptors in the dendrite membrane. On the other hand, vomeronasal neurons project directly to the brain; therefore, they have to transform the chemical information carried by the pheromone molecules into action potentials traveling over afferent nerve fibers. This is a complex process that, for convenience, is subdivided into two functional steps: transduction, the series of events by which VNO neurons transform the pheromone signals into a change in membrane potential (receptor potential); and coding, the process by which the membrane depolarization generated during transduction is converted into trains of action potentials in the nerve fibers. Both transduction and coding rely on the presence of ion channels in VNO neurons. Electrophysiological studies have provided a considerable amount of information on the biophysical and pharmacological properties of these channels, as well as on their regulation by intracellular signals. Indeed, ion channels represent the point of arrival for a chain of events that starts with the activation of pheromone receptors. G proteins, enzymes, and second messengers play a key role in the signaling pathway linking pheromone receptors to ion channels. In this section, we will review the available data on the microphysiology of pheromone signaling in VNO neurons. By using such information, we will also provide a general picture of the transduction and coding processes that take place in vomeronasal neurons during pheromone detection.

Electrophysiology of VNO neurons

Patch-clamp recordings have provided information on the electrical properties of rodent VNO neurons. These sensory neurons fire action potentials either in response to current injections or spontaneously (Holy et al. 2000; Inamura et al. 1997b; Leinders-Zufall et al. 2000; Liman and Corey 1996; Stowers et al. 2002; Tirindelli et al. 1998) and possess several voltage-gated ion channels, including TTX-sensitive Na^+ channels, delayed rectifier-type K^+ channels, and T- and L-type Ca^{2+} channels (Fieni et al. 2003; Ghiaroni et al. 2003; Inamura et al. 1997b; Liman and Corey 1996; Trotier et al. 1998). VNO neurons are highly excitable, as indicated by the observations that they can fire action potentials following current injection of just a few picoamperes (e.g., Liman and Corey 1996). This remarkable sensitivity is likely due to their high input resistance, which is related to a high membrane resistance in resting conditions (Fieni et al. 2003; Ghiaroni et al. 2003; Inamura et al. 1997b; Liman and Corey 1996). Although a similar sensitivity has been shown also for olfactory neurons (Lynch and Barry 1989; Maue and Dionne 1987), VNO neurons have unique firing properties in that they can fire tonically without adaptation during prolonged current injection of as little as 1 pA, whereas olfactory neurons required higher current levels (Liman and Corey 1996). Liman and Corey (1996) suggested that firing, by allowing temporal integration at the first central synapse, may represent a mechanism to enhance the sensitivity of the pheromone-detecting system, which cannot rely on a large convergence of sensory neurons in the accessory olfactory bulb as is the case for the olfactory neurons (Mombaerts 1999; Mori et al. 1999; Ressler et al. 1994; Vassar et al. 1994). By using a flat array of 61 extracellular electrodes, Holy et al. (2000) demonstrated that the firing response of VNO neurons to urine also showed little or no adaptation to prolonged stimulation.

Although early studies did not address the issue of a possible electrophysiological diversification across VNO neurons, recent data indicate that the relative expression of voltage-gated channels differs considerably between the two neuronal populations of the VNO. Apical neurons exhibit a large voltage-gated, TTX-sensitive Na^+ current, whereas the opposite is found in basal neurons, irrespective of cell membrane extension (Fieni et al. 2003). Since Na^+ current density affects the membrane excitability (for example, by setting the firing threshold), the difference in current density between apical and basal neurons suggests that these two neuronal populations exhibit different capabilities as to the generation of action potentials. Consistent with electrophysiological findings, in situ hybridization analysis of VNO sections revealed a gradient of expression for $Na_V1.3$ mRNA, with a high level in apical neurons and lower level in basal ones (Fieni et al. 2003). $Na_V1.3$ is the sodium channel α-subunit normally found primarily in neuronal cell bodies (Rasband and Shrager 2000).

In addition to sodium currents, vomeronasal neurons differ also in the transient (T-type), voltage-gated Ca^{2+} currents (Fieni et al. 2003). Specifically, magnitude of T-type current was larger in basal neurons than apical neurons. Since T-type channels are preferentially localized to dendrite in central neurons (Perez-Reyes 2003), it is conceivable that the difference in T-type currents between apical and basal neurons could be related to the length of their dendrites. T-type channels in the dendrite may play a role in sustaining the transmission of membrane depolarization from the microvilli membrane toward the cell body membrane, a mechanism of signal amplification that seems to operate in the dendritic tree of central neurons (Destexhe et al. 1998; Magee and Johnston 1995; Markram and Sakmann 1994; Pouille et al. 2000). Interestingly, olfactory neurons in rat do not possess T-type Ca^{2+} currents (Trombley and Westbrook 1991). As Liman and Corey (1996) pointed out, "the presence of T-type current in VNO and not in olfactory neurons indicates that it may play a role in defining the distinct electrical properties of VNO neurons."

Do the electrical properties of VNO neurons differ according to gender (sexual dichotomy)? This is an important issue for several reasons. First, sexual dimorphism occurs in the rodent vomeronasal system at multiple levels along its projection pathway (Keverne 2002); second, differences in the biophysical and pharmacological properties of voltage-gated currents between male and female VNO neurons have been documented in non-mammalian species (Fadool et al. 2001). A recent electrophysiological study has analyzed voltage-gated currents in VNO neurons of two mouse strains (BALB/c and CBA) and correlated them to sex in each strain (Dean et al. 2004). Slight differences were observed with CBA mice but not with the BALB/c strain. Thus, the authors concluded that mouse VNO neurons do not exhibit a clear-cut sexual dichotomy in their voltage-dependent membrane properties.

Transduction channel

Action potential firing mediated by voltage-gated ion channels represents the last step of a series of events that starts with the activation of pheromone receptors in the dendrite of VNO neurons. How is pheromonal information transferred from membrane receptors to voltage-gated channels? This fundamental issue has engaged researchers for several years and has come to a resolution only after the discovery of the transduction channel in VNO neurons, namely the TRPC2 channel (also known as TRP2 channel), a member of the transient receptor potential (TRP) family of ion channels first described in Drosophila phototransduction (Minke and Cook 2002). As mentioned above, transduction refers to the process by which pheromone receptor activation leads to the generation of a change in the membrane potential, namely a membrane depolarization. This is an essential step, for depolarization will eventually trigger opening of voltage-gated channels and therefore firing. The transduction channel is the ion channel that mediates the required membrane depolarization. In olfactory sensory neurons, cyclic nucleotide-gated (CNG) channels generate electrical signals during transduction (reviewed in Firestein 2001; Schild and Restrepo 1998). However, early studies established that VNO neurons did not possess CNG channels (Berghard et al. 1996; Liman and Corey 1996; Wu et al. 1996). By considering that key elements of the transduction pathway in the olfactory sensory neurons were not present in VNO neurons (Berghard and Buck 1996; Berghard et al. 1996; Wu et al. 1996), Liman and colleagues looked for other putative transduction channels and were able to detect the specific expression of TRPC2 channel transcript in rat VNO sensory neurons, but not in olfactory neurons or other neuronal and nonneuronal tissues (Liman et al. 1999). In situ hybridization with a digoxigenin-labeled rTRPC2 antisense RNA probe revealed that both apical and basal neurons expressed the TRPC2 channel transcript. Most importantly, they demonstrated with immunolabeling that the channel protein was restricted to the sensory microvilli of VNO neurons, which is the expected localization for a transduction channel. Ultrastructural evidence confirmed that the TRPC2 channel and the G protein subunits, $G\alpha_{i2}$ and $G\alpha_o$, were localized within the microvilli of VNO neurons (Menco et al. 2001). In agreement with in situ hybridization experiments by Liman et al. (1999), Menco and collaborators (2001) found that the microvilli of both apical and basal neurons contain TRPC2 channels. Subsequent behavioral studies with *TRPC2* knock-out mice further supported the notion that the TRPC2 channel is an essential component for pheromone detection (Leypold et al. 2002; Stowers et al. 2002). TRPC2$^{-/-}$ mice are characterized by a loss of both sex discrimination and male–male aggression, typical VNO-driven behavioral responses to conspecifics.

The electrophysiological properties and the mechanism of activation of TRPC2 channels in VNO neurons has been provided recently in an elegant study by Zufall's group (Lucas et al. 2003). The TRPC2 channel is a Ca^{2+}-permeable cation channel that is gated directly by diacylglycerol (DAG), a lipid messenger molecule that is produced in VNO neurons during pheromonal signaling. Consistently with previous molecular and behavioral findings (Leypold et al. 2002; Liman et al. 1999; Stowers et al. 2002), TRPC2 channels mediate ion currents underlying depolarizing receptor potential induced by application of dilute urine to single VNO neurons (Lucas et al. 2003). This cation channel is defective in *TRPC2$^{-/-}$* VNO neurons (Lucas et al. 2003). Thus, the TRPC2 channel is the transduction channel that links intracellular signaling following pheromone receptor activation to action potential firing in VNO neurons.

An open issue is the impact of Ca^{2+} entry mediated by TRPC2 channels in pheromonal transduction. In this regard it is important to recall that a Ca^{2+}-activated cation (CaNS) channel has been reported in VNO neurons (Liman 2003). This channel could be opened by calcium ions leaking into the VNO neuron microvilli through TRPC2 channels, and therefore it could enhance the receptor potential by providing an extra membrane depolarization. Whether this mechanism for signal amplification actually takes place in VNO neurons in situ awaits further experimentation. It is conceivable to expect a strong impact of the ionic composition of fluid bathing the microvilli in the actual response of VNO neurons, especially if one considers that a direct influx of urine into the VNO lumen can occur with licking (Wysocki et al. 1980).

As already reported in this review, recent electrophysiological findings indicate a differential density level for voltage-gated Na^+ and Ca^{2+} currents between apical and basal neurons (Fieni et al. 2003). According to in situ hybridization and immunocytochemistry findings, the TRPC2 channel is found both in apical and basal neurons (Liman et al. 1999). However, it is not know whether the functional properties of TRPC2 channels are similar in the two neuronal subpopulations. It would be also interesting to evaluate whether both apical and basal neurons express CaNS channels, since this channel could be involved in amplifying the primary sensory response mediated by TRPC2 channels.

TRPC2 is a pseudogene in humans, as it contains six deleterious mutations that generate premature stop codons (Liman and Innan 2003). By analyzing genomic DNA from 15 extant primate species, Liman and Innan also found that these mutations in the *TRPC2* gene began to accumulate in the common ancestor of Old World monkeys (e.g., rhesus and mandrill) and apes (e.g., orangutan and gorilla), many of which lack a VNO. On the other hand, *TRPC2* gene has been retained in prosimians (lemur, which has a functioning VNO) and in New World monkeys (e.g., squirrel monkey and marmoset). In these latter species, however, Liman and Innan (2003) found that selective pressure on the TRPC2 gene was relaxed, suggesting that in these primates VNO signaling may be vestigial or redundant. Consistent with the key role in pheromone transduction, the TRPC2 gene is under conspicuous selective pressure in the rodent lineage (Liman and Innan 2003).

Signaling cascade

The TRPC2 channel represents the final point of the signal transduction cascade initiated by the activation of pheromone receptor. The communication between these two membrane proteins is obtained through a complex intracellular pathway involving several other proteins (signaling cascade). Pheromone receptors (V1Rs and V2Rs) and several specific subunits of heterotrimeric G proteins have been identified in vomeronasal neurons, including $G\alpha_{i2}$, $G\alpha_o$, $G\alpha_{q/11}$, $G\beta_2$, $G\gamma_2$, and $G\gamma_8$ (Berghard et al. 1996; Jia and Halpern 1996; Runnenburger et al. 2002; Tirindelli and Ryba 1996; Wekesa et al. 2003). Immunocytochemistry and ultrastructural studies on $G\alpha_{i2}$, $G\alpha_o$, and $G\alpha_{q/11}$ have demonstrated their specific cellular localization in VNO microvilli (Berghard and Buck 1996; Liman et al. 1999; Menco et al. 2001; Wekesa et al. 2003), consistent with a potential role in signal transduction. As described above, the distribution pattern of these proteins across the VNO neuroepithelium has revealed that $G\alpha_{i2}$ is abundantly expressed in apical neurons whereas $G\alpha_o$ is confined to basal neurons (e.g., Berghard and Buck 1996; Fig. 3). Interestingly, their pattern of expression among neurons is closely matched by the expression patterns of V1Rs and V2Rs, respectively (Dulac and Axel 1995; Herrada and Dulac 1997; Matsunami and Buck 1997; Pantages and Dulac 2000; Rodriguez et al. 2002; Ryba and Tirindelli 1997). The spatial expression profile of identified $G\gamma$ subtypes revealed that $G\gamma_8$ is expressed preferentially by basal neurons (Runnenburger et al. 2002; Tirindelli and Ryba 1996), whereas apical neurons seem to express $G\gamma_2$ (Runnenburger et al. 2002). Finally, both types of neurons share a $G\beta_2$ subunit (Runnenburger et al. 2002) and probably $G\alpha_{q/11}$ (Wekesa et al. 2003). Thus, the early events in pheromone detection seem to rely on distinct receptor–G protein complexes ("V1R–$G\alpha_{i2}$–$G\gamma_2$" and "V2R–$G\alpha_o$–$G\gamma_8$") segregated into separate neuronal subsets (apical and basal neurons, respectively). This molecular-anatomical segregation might set the basis for quality coding in the VNO.

Although we have a lot of information on the molecular entities (receptors and G protein subunits) that may be involved in the pheromonal signaling, our knowledge on how these different molecules actually interact in a single VNO neuron is very poor. Studies on membrane preparations (Krieger et al. 1999; Wekesa and Anholt 1997; Wekesa et al. 2003) as well as electrophysiological observations (Holy et al. 2000; Inamura et al. 1997a; Lucas et al. 2003) indicate that the phospholipase C-β (PLC-β) signaling pathway plays a major role in the pheromone transduction. Biochemical data suggest that the G$\beta\gamma$ complex of pertussis toxin-sensitive G_i/G_o proteins mediates PLC activation in the VNO (Runnenburger et al. 2002). Also a recent study on microvillar membrane from female murine VNO has shown that blocking the G protein α-subunit of G_{i2} and G_o by adenosine diphosphate (ADP)-ribosylation does not block PLC-mediated signaling induced by stimulation with male urine (Wekesa et al. 2003). Given the different spatial expression profile of Gγ subunits across the VNO (Gγ_2 in apical neurons and Gγ_8 in basal neurons), it has been suggested that PLC activation in the two populations of vomeronasal neurons might be mediated by different G$\beta\gamma$ complexes (Runnenburger et al. 2002).

If we accept the notion that the $\beta\gamma$ complex transmits pheromonal information to vomeronasal PLC, then an obvious question is about the role of the α-subunits in the signaling cascade. A number of studies have shown that G protein α-subunits play a role in modulating voltage-gated Ca^{2+} currents (Dolphin 2003), which are expressed by VNO neurons (Fieni et al. 2003; Liman and Corey 1996). Furthermore, recent evidence indicates that Gα_i regulates the activity of the G protein-activated inwardly rectifying K^+ (GIRK) channels (Peleg et al. 2002). An inward-rectifier potassium current has been described in mouse VNO neurons (Liman and Corey 1996). Since inward rectifier K^+ channels play a role in controlling the resting potential, any change in their activity might have an impact on the membrane excitability. Thus, α-subunits of G_{i2}/G_o might concur in defining the profile of electrical signaling for each activated VNO neurons and, therefore, in signal coding.

How activation of PLC signaling affects the electrical excitability of vomeronasal neurons has become clear recently with the discovery that the transduction TRPC2 channel is directly gated by DAG (Lucas et al. 2003). DAG is a lipid messenger produced by the PLC-mediated hydrolysis of phosphatidyl inositol bisphosphate (PIP_2), a membrane lipid constituent. In addition to DAG formation, however, PIP_2 hydrolysis gives rise to another second messenger, inositol-1,4,5-triphosphate (IP_3). Several studies have shown that IP_3 level increases in the VNO membrane preparations during stimulation with urinary pheromones (Kroner et al. 1996; Rossler et al. 2000; Runnenburger et al. 2002; Sasaki et al. 1999; Wekesa and Anholt 1997; Wekesa et al. 2003). Immunohistochemical analysis suggests that the type III IP_3 receptor occurs throughout the sensory zone of the rat VNO (Brann et al. 2002). Injection of IP_3 into rat VNO neurons induces inward current responses under whole-cell voltage clamp conditions (Inamura et al. 1997b). Finally, action potential firing induced by urine preparation in rat VNO neurons is blocked by ruthenium red, an IP_3 receptor inhibitor (Inamura et al. 1997a). As a whole, these findings support the notion that IP_3 may play a role in the early events of pheromone transduction. On the other hand, the careful and extensive patch-clamp study of sensory current in mouse VNO neurons by Lucas et al. (2003) led to the conclusion that IP_3 does not represent a primary step for activation of the transduction conductance. Although they were able to obtain inward current upon cell dialysis with IP_3, the kinetic, ionic, and pharmacological profile of this current did not match that evoked by pheromone. It is therefore possible that IP_3 signaling could play a role in events downstream of the activation of transduction TRPC2 channels. A similar argument may hold also for the involvement of arachidonic acid or other polyunsaturated fatty acids in pheromone transduction (Lucas et al. 2003; Spehr et al. 2002).

Electrophysiological experiments (Holy et al. 2000; Lucas et al. 2003) identified PCL-β as a key element of the signaling cascade in mammalian VNO neurons and also confirmed previous observations that cyclic nucleotides were not essential in sensory transduction. In patch-clamp experiments, dialysis of VNO neurons with cyclic AMP (cAMP) failed to induce any membrane currents both in mouse (Liman and Corey 1996) and rat (Sasaki et al. 1999). Nevertheless, there is evidence that suggests possible involvement of cAMP in the activity of VNO neurons. A high level of adenyl cyclase type II (ACII) was found in all VNO neurons (Berghard and Buck 1996). More recently, Rössler et al. (2000) found that another AC isoform, ACVI, was expressed by a subset of cells across the whole sensory epithelium of VNO. Interestingly, $G\alpha_{i2}$, but not $G\alpha_o$, is able to inhibit adenyl cyclases (Hurley 1999). Data from an in vitro cAMP accumulation assay with mouse VNO membranes indicate that cAMP levels decreases after the exposure to urine-derived compounds (Zhou and Moss 1997). A decrease in cAMP level was also observed by Rössler and collaborators (2000) in rat VNO microvillar preparations stimulated with urinary components. However, they found evidence that the decrease was due to the activation of the phosphoinositol pathway rather than to enhancement of phosphodiesterase activity or to an inhibition of AC via G proteins. Specifically, they claim that a calcium-inhibited and protein kinase C-inhibited ACVI localized in the microvillar preparations seems to be responsible for the decrease in cAMP level during pheromonal stimulation involving either lipophilic urinary components, which stimulate PCLβ via G_i protein in the VNO (Krieger et al. 1999), or α_{2u}, a pheromonal urinary component that activates PLCβ via G_o protein (Krieger et al. 1999). Moreover, Rössler et al. (2000) showed also that pretreatment of VNO microvillar preparations with ACII antibodies did not affect the decrease in cAMP level, whereas pretreatment with ACVI specific antibodies did in a dose-dependent manner. The evaluation of the kinetics of the second messenger responses in the subsecond time range revealed that the reduction of the cAMP concentration was delayed compared to the increase in IP$_3$ level, further suggesting that cAMP was not a primary messenger in the chemo-electrical transduction in rodent VNO and that could be involved in regulating the responsiveness of VNO neurons (Rossler et al. 2000). Clearly, further experiments are required to elucidate the role of cAMP in the pheromone signaling cascade.

In summary, the available information on the molecular, biochemical, and electrophysiological properties of vomeronasal neurons has allowed us to gain a better insight into the mechanisms involved in pheromone reception. Figure 5 shows a schematic drawing of the main events that characterize pheromone transduction, namely the transformation of a chemical signal (pheromone molecule) into an electrical signal (receptor potential). This activity represents the prerequisite to relay sensory information to the accessory olfactory bulb through the axons of VNO neurons. Pheromonal information is transmitted in the form of a series of action potentials, which are initiated by the activity of voltage-gated channels in VNO neurons.

Coding of pheromonal signals

The outcome of pheromone transduction is a membrane depolarization in VNO neurons (e.g., Lucas et al. 2003), which can either initiate action potential firing if the neuron is silent, or modulate an ongoing impulse discharge (Holy et al. 2000; Inamura et al. 1999; Leinders-Zufall et al. 2000; Lucas et al. 2003; Stowers et al. 2002). This functional step in sensory

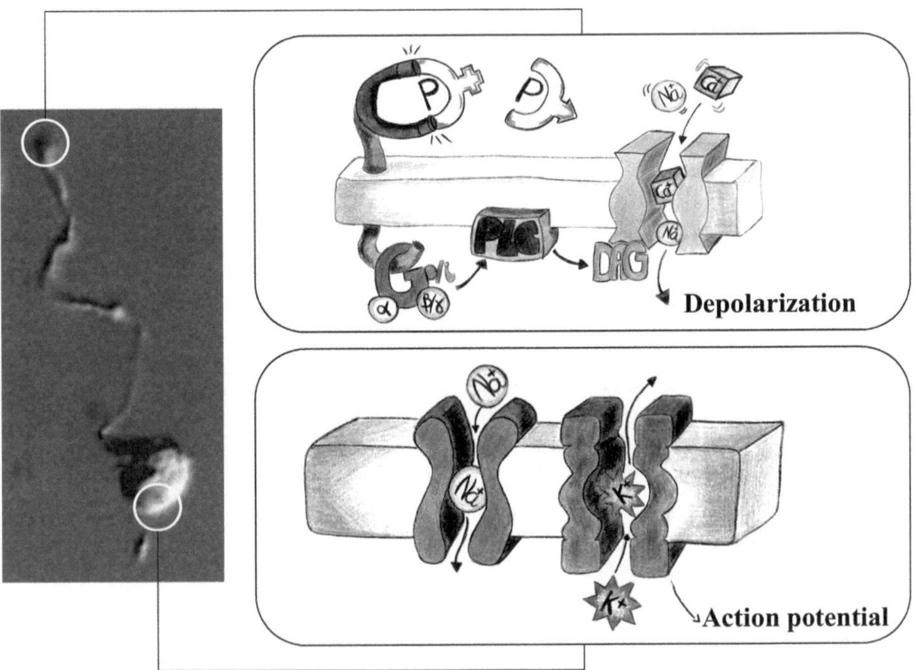

Fig. 5 Pheromone transduction pathway and electrical signaling in vomeronasal neurons. Early events in pheromone detection occur in the microvilli at end of the dendrite, which protrude inside the VNO lumen. At this level, binding of pheromone molecules (*P*) with their specific membrane receptors activates an intracellular signaling cascade involving G proteins (G_{i2} in the apical neurons, G_o in basal neurons), phospholipase C-β (PLC) and the second messenger, diacylglycerol (DAG). DAG is the gating factor for a cation channel, TRPC2. Opening of this channel leads to an influx of Na^+/Ca^{2+} that causes a membrane depolarization, the so-called receptor potential. Membrane depolarization spreads along the dendrite and activate voltage-gated Na^+ and K^+ channels with consequent generation of action potentials. Firing is an essential step in the overall process of pheromone detection: impulse propagation along the nerve fibers allows VNO to communicate with the first central processing unit, namely the accessory olfactory bulb. Activation of PLC is thought to be mediated by the βγ complex of G proteins. The role of the α-subunit is not known. A possibility is that it might modulate the activity of other ion channels in the neuronal membrane

reception is referred to as coding. Within the context of this definition, coding provides brain with information about intensity (concentration) of the stimulus (intensity coding=frequency of action potentials) and the timing of stimulus occurrence (temporal coding=the duration of action potential firing). Both intensity and temporal coding have been demonstrated for mouse and rat VNO neurons (Holy et al. 2000; Inamura et al. 1999; Leinders-Zufall et al. 2000; Stowers et al. 2002). There are other aspects of sensory coding, however, that rely on the organization and activity of a population of neurons rather than on the activity of a single cell. Quality coding to monitor differences in a stimulus's chemical species is one such aspect. There are indications that in the VNO, quality coding might rely on the segregation of molecular and functional properties in the two separate neuronal subpopulations (apical and basal neurons). An in vitro study has provided evidence that in rat VNO, $G\alpha_{i2}$ is activated by hydrophobic molecules extracted from male urine, whereas $G\alpha_o$ is activated upon stimulation with protein pheromones (Krieger et al. 1999). Each G protein subtype is stereotypically co-expressed with one of the candidate pheromone receptors (V1R or V2R) and in separate neuronal populations (apical and basal neurons). Thus, findings by Krieger et al. (1999) support the concept that anatomical and molecular segregation of sensory neurons may underlie the discrimination between hydrophobic compounds and protein pheromones (quality coding). Moreover, apical and basal neurons project to different area of the AOB (Halpern et al. 1998; Jia and Halpern 1996; Wekesa and Anholt 1999).

Electrophysiological evidence for anatomically-based quality coding in the VNO has been provided by Inamura et al. (1999). They used the cell-attached patch clamp method to record impulse activity from sensory neurons in vomeronasal slices of female Wistar rats following application of different urine preparations. They found that there are at least two segregated neuronal laminae that have different pheromonal sensitivities: The majority of neurons responding to conspecific male rat urine was located in the $G\alpha_{i2}$ layer, whereas the majority of neurons responding to conspecific female rat urine and male urine from a different rat strain (Donryu or Sprague-Dawley) localized to the $G\alpha_o$ layer.

Pheromonal receptors (V1Rs and V2Rs) appear to be expressed differentially in small subsets of VNO neurons. Although we still lack information about the chemical identity of the natural ligands for VRs (with the exception of V1Rb2: Boschat et al. 2002), and therefore about the breadth of VRs responsiveness, Leinders-Zufall et al. (2000) demonstrated with an optical imaging approach that mouse VNO neurons are tuned to recognize only one or perhaps very few pheromonal components. In addition, increasing concentrations of ligands do not induce recruitment of other sensory neurons, as is the case for olfactory neurons that use a combinatorial receptor coding (Duchamp-Viret et al. 1999; Firestein 2001; Malnic et al. 1999). In other words, a given ligand activates a unique, non-overlapping subset of VNO neurons, and this activation is concentration-invariant (Leinders-Zufall et al. 2000).

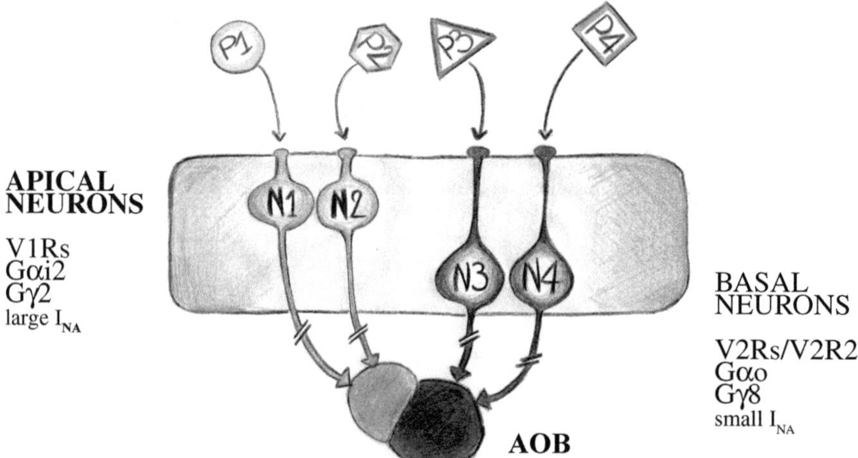

Fig. 6 The anatomical, molecular, and electrophysiological basis for sensory coding in mammalian vomeronasal organ (VNO). Sensory neurons in VNO segregate into two main layers: apical neurons endowed with a short dendrite, and basal neurons endowed with a long dendrite. Apical and basal neurons project to different areas of the accessory olfactory bulb (*AOB*). This anatomical distinction also reflects differences in the set of transduction molecules (such as pheromone receptors and G protein subunits) and in the electrophysiological properties (such as the density of voltage-gated sodium current, I_{Na}) in sensory neurons. Thus, two main separate signaling routes exist in VNO for pheromone detection (spatial coding). In addition, in each of these signaling routes, small subsets of neurons (*N1, N2, N3, N4*) are tuned to detect only a given chemical (*P1, P2, P3, P4*) of the pheromonal blend. Thus, signal transfer in the VNO appears to be organized according to labeled lines, whereas quality code likely relies on the patterns of activity across the neuronal populations. Note that although each vomeronasal receptor (either V1R or V2R) appears to be expressed in a small subset of vomeronasal neurons, $V2R_2$ is found in all basal neurons

These findings strongly suggest the existence of separate transduction routes in the VNO that are tuned for the detection of a specific pheromonal component. Luo and Katz (2004) have proposed that pheromonal signals might be encoded by labeled lines in the VNO ("one cell–one receptor–one ligand" model). Inamura et al. (1999) also showed that most single vomeronasal neurons responded to only one of the urine preparations, consistent with a labeled-line strategy proposed by Luo and Katz (2004).

On the other hand, it is likely that the VNO is challenged by a blend of active molecules rather than by a single, pure pheromone. If each neuron responds to only one component of the urine, then simultaneous firing of a population of neurons responsive to different components of the pheromonal blend may be necessary to obtain the appropriate central representation to drive the behavioral response. Thus, an across-neuron pattern strategy might underlie the actual information transferred to the AOB. Direct evidence that sensory codes in the VNO might involve patterns of activity across the neuronal population has been provided by imaging (Leinders-Zufall et al. 2000) and electrophysiological studies (Holy et al. 2000).

A scheme summarizing the available information about the sensory coding at the level of the VNO is provided in Fig. 6. Coding occurs also at the AOB level, that is, the first central processing unit for pheromonal information. Further details can be found in recent reviews (Dulac 2000; Luo and Katz 2004).

Concluding remarks

The knowledge of the structure and functions of the vomeronasal organ has been greatly advanced in the last few years. The nature of the stimuli that can excite the vomeronasal receptor neurons is being unraveled. The major boost was achieved with the discovery of the different families of vomeronasal receptors. Other major improvements were made in the knowledge of the intracellular signaling and in the electrophysiological responses to the stimulation. However, several features of the coupling between chemical and electrical responses remain to be elucidated to completely decipher the "Rosetta stone" of vomeronasal chemical signaling.

Acknowledgements. This work was supported by the Italian Board of Education and University (COFIN and FIRB).

References

Abel EL (1991) Alarm substance emitted by rats in the forced-swim test is a low volatile pheromone. Physiol Behav 50:723–727

Adler E, Hoon MA, Mueller KL, Chandrashekar J, Ryba NJ, Zuker CS (2000) A novel family of mammalian taste receptors. Cell 100:693–702

Aungst JL, Heyward PM, Puche AC, Karnup SV, Hayar A, Szabo G, Shipley MT (2003) Centre-surround inhibition among olfactory bulb glomeruli. Nature 426:623–629

Bacchini A, Gaetani E, Cavaggioni A (1992) Pheromone binding proteins of the mouse, Mus musculus. Experientia 48:419–421

Belluscio L, Koentges G, Axel R, Dulac C (1999) A map of pheromone receptor activation in the mammalian brain. Cell 97:209–220

Benton D (1982) The influence of androstenol—a putative human pheromone—on mood throughout the menstrual cycle. Biol Psychol 15:249–256

Berghard A, Buck LB (1996) Sensory transduction in vomeronasal neurons: evidence for G alpha o, G alpha i2, and adenylyl cyclase II as major components of a pheromone signaling cascade. J Neurosci 16:909–918

Berghard A, Buck LB, Liman ER (1996) Evidence for distinct signaling mechanisms in two mammalian olfactory sense organs. Proc Natl Acad Sci U S A 93:2365–2369

Bertmar G (1981) Evolution of vomeronasal organs in vertebrates. Evolution 35:359–366

Bhatnagar KP, Smith TD (2003) The human vomeronasal organ. V. An interpretation of its discovery by Ruysch, Jacobson, or Kolliker, with an English translation of Kolliker (1877). Anat Rec 270B:4–15

Bocskei Z, Groom CR, Flower DR, Wright CE, Phillips SE, Cavaggioni A, Findlay JB, North AC (1992) Pheromone binding to two rodent urinary proteins revealed by X-ray crystallography. Nature 360:186–188

Booth KK, Katz LS (2000) Role of the vomeronasal organ in neonatal offspring recognition in sheep. Biol Reprod 63:953–958

Boschat C, Pelofi C, Randin O, Roppolo D, Luscher C, Broillet MC, Rodriguez I (2002) Pheromone detection mediated by a V1r vomeronasal receptor. Nat Neurosci 5:1261–1262

Brann JH, Dennis JC, Morrison EE, Fadool DA (2002) Type-specific inositol 1,4,5-trisphosphate receptor localization in the vomeronasal organ and its interaction with a transient receptor potential channel, TRPC2. J Neurochem 83:1452–1460

Brennan PA, Schellinck HM, Keverne EB (1999) Patterns of expression of the immediate-early gene egr-1 in the accessory olfactory bulb of female mice exposed to pheromonal constituents of male urine. Neuroscience 90:1463–1470

Brouette-Lahlou I, Godinot F, Vernet-Maury E (1999) The mother rat's vomeronasal organ is involved in detection of dodecyl propionate, the pup's preputial gland pheromone. Physiol Behav 66:427–436

Brown EM, Gamba G, Riccardi D, Lombardi M, Butters R, Kifor O, Sun A, Hediger MA, Lytton J, Hebert SC (1993) Cloning and characterization of an extracellular Ca(2+)-sensing receptor from bovine parathyroid. Nature 366:575–580

Brown W, Eisner T, Whittaker R (1970) Allomones and kairomones: transspecific chemical messengers. Bio-Science 20:21–22

Bruce HM (1960) A block to pregnancy in the mouse caused by proximity of strange males. J Reprod Fertil 1:96–103

Buck L, Axel R (1991) A novel multigene family may encode odorant receptors: a molecular basis for odor recognition. Cell 65:175–187

Buck LB (1996) Information coding in the vertebrate olfactory system. Annu Rev Neurosci 19:517–544

Cao Y, Oh BC, Stryer L (1998) Cloning and localization of two multigene receptor families in goldfish olfactory epithelium. Proc Natl Acad Sci U S A 95:11987–11992

Cavaggioni A, Mucignat-Caretta C (2000) Major urinary proteins, alpha(2U)-globulins and aphrodisin. Biochim Biophys Acta 1482:218–228

Cavaggioni A, Findlay JB, Tirindelli R (1990) Ligand binding characteristics of homologous rat and mouse urinary proteins and pyrazine-binding protein of calf. Comp Biochem Physiol B 96:513–520

Cavaggioni A, Mucignat C, Tirindelli R (1999) Pheromone signalling in the mouse: role of urinary proteins and vomeronasal organ. Arch Ital Biol 137:193–200

Cavaggioni A, Mucignat-Caretta C, Zagotto G (2003) Absolute configuration of 2-sec-butyl-4,5-dihydrothiazole in male mouse urine. Chem Senses 28:791–797

Chess A, Simon I, Cedar H, Axel R (1994) Allelic inactivation regulates olfactory receptor gene expression. Cell 78:823–834

Clark AJ, Ghazal P, Bingham RW, Barrett D, Bishop JO (1985) Sequence structures of a mouse major urinary protein gene and pseudogene compared. Embo J 4:3159–3165

Comfort A (1971) Likelihood of human pheromones. Nature 230:432–3 passim

Conigrave AD, Quinn SJ, Brown EM (2000) L-amino acid sensing by the extracellular Ca2+-sensing receptor. Proc Natl Acad Sci U S A 97:4814–4819

Cowan CM, Roskams AJ (2002) Apoptosis in the mature and developing olfactory neuroepithelium. Microsc Res Tech 58:204–215

Darwish Marie A, Veggerby C, Robertson DH, Gaskell SJ, Hubbard SJ, Martinsen L, Hurst JL, Beynon RJ (2001) Effect of polymorphisms on ligand binding by mouse major urinary proteins. Protein Sci 10:411–417

Dean DM, Mazzatenta A, Menini A (2004) Voltage-activated current properties of male and female mouse vomeronasal sensory neurons: sexually dichotomous? J Comp Physiol A Neuroethol Sens Neural Behav Physiol 190:491–499

Del Punta K, Rothman A, Rodriguez I, Mombaerts P (2000) Sequence diversity and genomic organization of vomeronasal receptor genes in the mouse. Genome Res 10:1958–1967

Del Punta K, Leinders-Zufall T, Rodriguez I, Jukam D, Wysocki CJ, Ogawa S, Zufall F, Mombaerts P (2002a) Deficient pheromone responses in mice lacking a cluster of vomeronasal receptor genes. Nature 419:70–74

Del Punta K, Puche A, Adams NC, Rodriguez I, Mombaerts P (2002b) A divergent pattern of sensory axonal projections is rendered convergent by second-order neurons in the accessory olfactory bulb. Neuron 35:1057–1066

Destexhe A, Neubig M, Ulrich D, Huguenard J (1998) Dendritic low-threshold calcium currents in thalamic relay cells. J Neurosci 18:3574–3588

Dolphin AC (2003) G protein modulation of voltage-gated calcium channels. Pharmacol Rev 55:607–627

Dorries KM, Adkins-Regan E, Halpern BP (1997) Sensitivity and behavioral responses to the pheromone androstenone are not mediated by the vomeronasal organ in domestic pigs. Brain Behav Evol 49:53–62

Doty RL, Dunbar I (1974) Attraction of beagles to conspecific urine, vaginal and anal sac secretion odors. Physiol Behav 12:825–833

Doving KB, Trotier D (1998) Structure and function of the vomeronasal organ. J Exp Biol 201:2913–2925

Duchamp-Viret P, Chaput MA, Duchamp A (1999) Odor response properties of rat olfactory receptor neurons. Science 284:2171–2174

Dudley CA, Moss RL (1999) Activation of an anatomically distinct subpopulation of accessory olfactory bulb neurons by chemosensory stimulation. Neuroscience 91:1549–1556

Dulac C (2000) Sensory coding of pheromone signals in mammals. Curr Opin Neurobiol 10:511–518

Dulac C, Axel R (1995) A novel family of genes encoding putative pheromone receptors in mammals. Cell 83:195–206

Emes RD, Beatson SA, Ponting CP, Goodstadt L (2004) Evolution and comparative genomics of odorant- and pheromone-associated genes in rodents. Genome Res 14:591–602

Fadem BH (1987) Activation of estrus by pheromones in a marsupial: stimulus control and endocrine factors. Biol Reprod 36:328–332

Fadool DA, Wachowiak M, Brann JH (2001) Patch-clamp analysis of voltage-activated and chemically activated currents in the vomeronasal organ of Sternotherus odoratus (stinkpot/musk turtle). J Exp Biol 204:4199–4212

Feinstein P, Mombaerts P (2004) A contextual model for axonal sorting into glomeruli in the mouse olfactory system. Cell 117:817–831

Feinstein P, Bozza T, Rodriguez I, Vassalli A, Mombaerts P (2004) Axon guidance of mouse olfactory sensory neurons by odorant receptors and the beta2 adrenergic receptor. Cell 117:833–846

Fernandez-Fewell GD, Meredith M (1994) c-fos expression in vomeronasal pathways of mated or pheromone-stimulated male golden hamsters: contributions from vomeronasal sensory input and expression related to mating performance. J Neurosci 14:3643–3654

Fieni F, Ghiaroni V, Tirindelli R, Pietra P, Bigiani A (2003) Apical and basal neurones isolated from the mouse vomeronasal organ differ for voltage-dependent currents. J Physiol 552:425–436

Firestein S (2001) How the olfactory system makes sense of scents. Nature 413:211–218

Getchell TV (1986) Functional properties of vertebrate olfactory receptor neurons. Physiol Rev 66:772–818

Ghiaroni V, Fieni F, Tirindelli R, Pietra P, Bigiani A (2003) Ion conductances in supporting cells isolated from the mouse vomeronasal organ. J Neurophysiol 89:118–127

Giorgi D, Friedman C, Trask BJ, Rouquier S (2000) Characterization of nonfunctional V1R-like pheromone receptor sequences in human. Genome Res 10:1979–1985

Glusman G, Yanai I, Rubin I, Lancet D (2001) The complete human olfactory subgenome. Genome Res 11:685–702

Goodwin M, Gooding KM, Regnier F (1979) Sex pheromone in the dog. Science 203:559–561

Guillamon A, Segovia S (1997) Sex differences in the vomeronasal system. Brain Res Bull 44:377–382

Gulyas B, Keri S, O'Sullivan BT, Decety J, Roland PE (2004) The putative pheromone androstadienone activates cortical fields in the human brain related to social cognition. Neurochem Int 44:595–600

Halem HA, Cherry JA, Baum MJ (1999) Vomeronasal neuroepithelium and forebrain Fos responses to male pheromones in male and female mice. J Neurobiol 39:249–263

Halpern M (1987) The organization and function of the vomeronasal system. Annu Rev Neurosci 10:325–362

Halpern M, Martinez-Marcos A (2003) Structure and function of the vomeronasal system: an update. Prog Neurobiol 70:245–318

Halpern M, Jia C, Shapiro LS (1998) Segregated pathways in the vomeronasal system. Microsc Res Tech 41:519–529

Hegde AN (2003) MHC molecules in the vomeronasal organ: contributors to pheromonal discrimination? Trends Neurosci 26:646–650

Herrada G, Dulac C (1997) A novel family of putative pheromone receptors in mammals with a topographically organized and sexually dimorphic distribution. Cell 90:763–773

Hildebrand JG (1995) Analysis of chemical signals by nervous systems. Proc Natl Acad Sci U S A 92:67–74
Holy TE, Dulac C, Meister M (2000) Responses of vomeronasal neurons to natural stimuli. Science 289:1569–1572
Hoon MA, Adler E, Lindemeier J, Battey JF, Ryba NJ, Zuker CS (1999) Putative mammalian taste receptors: a class of taste-specific GPCRs with distinct topographic selectivity. Cell 96:541–551
Hudson R, Distel H (1986) Pheromonal release of suckling in rabbits does not depend on the vomeronasal organ. Physiol Behav 37:123–128
Hurley JH (1999) Structure, mechanism, and regulation of mammalian adenylyl cyclase. J Biol Chem 274:7599–7602
Hurst JL, Payne CE, Nevison CM, Marie AD, Humphries RE, Robertson DH, Cavaggioni A, Beynon RJ (2001) Individual recognition in mice mediated by major urinary proteins. Nature 414:631–634
Inamura K, Kashiwayanagi M, Kurihara K (1997a) Blockage of urinary responses by inhibitors for IP3-mediated pathway in rat vomeronasal sensory neurons. Neurosci Lett 233:129–132
Inamura K, Kashiwayanagi M, Kurihara K (1997b) Inositol-1,4,5-trisphosphate induces responses in receptor neurons in rat vomeronasal sensory slices. Chem Senses 22:93–103
Inamura K, Matsumoto Y, Kashiwayanagi M, Kurihara K (1999) Laminar distribution of pheromone-receptive neurons in rat vomeronasal epithelium. J Physiol 517:731–739
Ishii T, Hirota J, Mombaerts P (2003) Combinatorial coexpression of neural and immune multigene families in mouse vomeronasal sensory neurons. Curr Biol 13:394–400
Isles AR, Baum MJ, Ma D, Keverne EB, Allen ND (2001) Urinary odour preferences in mice. Nature 409:783–784
Jacobson L, Trotier D, Doving KB (1998) Anatomical description of a new organ in the nose of domesticated animals by Ludvig Jacobson (1813). Chem Senses 23:743–754
Jahnke V, Merker HJ (2000) Electron microscopic and functional aspects of the human vomeronasal organ. Am J Rhinol 14:63–67
Jia C, Halpern M (1996) Subclasses of vomeronasal receptor neurons: differential expression of G proteins (Gi alpha 2 and G(o alpha)) and segregated projections to the accessory olfactory bulb. Brain Res 719:117–128
Jingami H, Nakanishi S, Morikawa K (2003) Structure of the metabotropic glutamate receptor. Curr Opin Neurobiol 13:271–278
Johns MA, Feder HH, Komisaruk BR, Mayer AD (1978) Urine-induced reflex ovulation in anovulatory rats may be a vomeronasal effect. Nature 272:446–448
Johnston RE (1998) Pheromones, the vomeronasal system, and communication. From hormonal responses to individual recognition. Ann N Y Acad Sci 855:333–348
Johnston RE, Peng M (2000) The vomeronasal organ is involved in discrimination of individual odors by males but not by females in golden hamsters. Physiol Behav 70:537–549
Jones DT, Reed RR (1989) Golf: an olfactory neuron specific-G protein involved in odorant signal transduction. Science 244:790–795
Karlson P, Lüscher M (1959) "Pheromones": a new term for a class of biologically active substances. Nature 183:55–56
Kaupmann K, Huggel K, Heid J, Flor PJ, Bischoff S, Mickel SJ, McMaster G, Angst C, Bittiger H, Froestl W, Bettler B (1997) Expression cloning of GABA(B) receptors uncovers similarity to metabotropic glutamate receptors. Nature 386:239–246
Keverne EB (1999) The vomeronasal organ. Science 286:716–720
Keverne EB (2002) Pheromones, vomeronasal function, and gender-specific behavior. Cell 108:735–738
Krautwurst D, Yau KW, Reed RR (1998) Identification of ligands for olfactory receptors by functional expression of a receptor library. Cell 95:917–926
Krieger J, Schmitt A, Lobel D, Gudermann T, Schultz G, Breer H, Boekhoff I (1999) Selective activation of G protein subtypes in the vomeronasal organ upon stimulation with urine-derived compounds. J Biol Chem 274:4655–4662
Kroner C, Breer H, Singer AG, O'Connell RJ (1996) Pheromone-induced second messenger signaling in the hamster vomeronasal organ. Neuroreport 7:2989–2992
Kuang D, Yao Y, Wang M, Pattabiraman N, Kotra LP, Hampson DR (2003) Molecular similarities in the ligand binding pockets of an odorant receptor and the metabotropic glutamate receptors. J Biol Chem 278:42551–42559
Kunishima N, Shimada Y, Tsuji Y, Sato T, Yamamoto M, Kumasaka T, Nakanishi S, Jingami H, Morikawa K (2000) Structural basis of glutamate recognition by a dimeric metabotropic glutamate receptor. Nature 407:971–977

Lane RP, Cutforth T, Axel R, Hood L, Trask BJ (2002) Sequence analysis of mouse vomeronasal receptor gene clusters reveals common promoter motifs and a history of recent expansion. Proc Natl Acad Sci U S A 99:291–296

Leinders-Zufall T, Lane AP, Puche AC, Ma W, Novotny MV, Shipley MT, Zufall F (2000) Ultrasensitive pheromone detection by mammalian vomeronasal neurons. Nature 405:792–796

Leon M, Moltz H (1973) Endocrine control of the maternal pheromone in the postpartum female rat. Physiol Behav 10:65–67

Leypold BG, Yu CR, Leinders-Zufall T, Kim MM, Zufall F, Axel R (2002) Altered sexual and social behaviors in trp2 mutant mice. Proc Natl Acad Sci U S A 99:6376–6381

Liman ER (2003) Regulation by voltage and adenine nucleotides of a Ca2+-activated cation channel from hamster vomeronasal sensory neurons. J Physiol 548:777–787

Liman ER, Corey DP (1996) Electrophysiological characterization of chemosensory neurons from the mouse vomeronasal organ. J Neurosci 16:4625–4637

Liman ER, Innan H (2003) Relaxed selective pressure on an essential component of pheromone transduction in primate evolution. Proc Natl Acad Sci U S A 100:3328–3332

Liman ER, Corey DP, Dulac C (1999) TRP2: a candidate transduction channel for mammalian pheromone sensory signaling. Proc Natl Acad Sci U S A 96:5791–5796

Loconto J, Papes F, Chang E, Stowers L, Jones EP, Takada T, Kumanovics A, Fischer Lindahl K, Dulac C (2003) Functional expression of murine V2R pheromone receptors involves selective association with the M10 and M1 families of MHC class Ib molecules. Cell 112:607–618

Lucas P, Ukhanov K, Leinders-Zufall T, Zufall F (2003) A diacylglycerol-gated cation channel in vomeronasal neuron dendrites is impaired in TRPC2 mutant mice: mechanism of pheromone transduction. Neuron 40:551–561

Luo M, Katz LC (2004) Encoding pheromonal signals in the mammalian vomeronasal system. Curr Opin Neurobiol 14:428–434

Luo M, Fee MS, Katz LC (2003) Encoding pheromonal signals in the accessory olfactory bulb of behaving mice. Science 299:1196–1201

Lynch JW, Barry PH (1989) Action potentials initiated by single channels opening in a small neuron (rat olfactory receptor). Biophys J 55:755–768

Magee JC, Johnston D (1995) Synaptic activation of voltage-gated channels in the dendrites of hippocampal pyramidal neurons. Science 268:301–304

Malnic B, Hirono J, Sato T, Buck LB (1999) Combinatorial receptor codes for odors. Cell 96:713–723

Markram H, Sakmann B (1994) Calcium transients in dendrites of neocortical neurons evoked by single subthreshold excitatory postsynaptic potentials via low-voltage-activated calcium channels. Proc Natl Acad Sci U S A 91:5207–5211

Martini S, Silvotti L, Shirazi A, Ryba NJ, Tirindelli R (2001) Co-expression of putative pheromone receptors in the sensory neurons of the vomeronasal organ. J Neurosci 21:843–848

Masera T (1943) Su l'esistenza di un particolare organo olfattivo nel setto nasale della cavia e di altri roditori. Arch Ital Anat Embriol 48:157–212

Matsunami H, Buck LB (1997) A multigene family encoding a diverse array of putative pheromone receptors in mammals. Cell 90:775–784

Matsunami H, Montmayeur JP, Buck LB (2000) A family of candidate taste receptors in human and mouse. Nature 404:601–604

Maue RA, Dionne VE (1987) Patch-clamp studies of isolated mouse olfactory receptor neurons. J Gen Physiol 90:95–125

Max M, Shanker YG, Huang L, Rong M, Liu Z, Campagne F, Weinstein H, Damak S, Margolskee RF (2001) Tas1r3, encoding a new candidate taste receptor, is allelic to the sweet responsiveness locus Sac. Nat Genet 28:58–63

McClintock MK (1984) Estrous synchrony: modulation of ovarian cycle length by female pheromones. Physiol Behav 32:701–705

McGlone JJ (1985) Olfactory cues and pig agonistic behavior: evidence for a submissive pheromone. Physiol Behav 34:195–198

Menco BP, Carr VM, Ezeh PI, Liman ER, Yankova MP (2001) Ultrastructural localization of G-proteins and the channel protein TRP2 to microvilli of rat vomeronasal receptor cells. J Comp Neurol 438:468–489

Meredith M (1998) Vomeronasal function. Chem Senses 23:463–466

Meredith M, Westberry JM (2004) Distinctive responses in the medial amygdala to same-species and different-species pheromones. J Neurosci 24:5719–5725

Michael RP, Keverne EB (1970) A male sex-attractant pheromone in rhesus monkey vaginal secretions. J Endocrinol 46:xx–xxi

Minke B, Cook B (2002) TRP channel proteins and signal transduction. Physiol Rev 82:429–472

Miyawaki A, Matsushita F, Ryo Y, Mikoshiba K (1994) Possible pheromone-carrier function of two lipocalin proteins in the vomeronasal organ. Embo J 13:5835–5842

Moltz H, Lee TM (1981) The maternal pheromone of the rat: identity and functional significance. Physiol Behav 26:301–306

Mombaerts P (1996) Targeting olfaction. Curr Opin Neurobiol 6:481–486

Mombaerts P (1999) Seven-transmembrane proteins as odorant and chemosensory receptors. Science 286:707–711

Mombaerts P, Wang F, Dulac C, Chao SK, Nemes A, Mendelsohn M, Edmondson J, Axel R (1996) Visualizing an olfactory sensory map. Cell 87:675–686

Monti-Bloch L, Jennings-White C, Berliner DL (1998) The human vomeronasal system. A review. Ann N Y Acad Sci 855:373–389

Montmayeur JP, Liberles SD, Matsunami H, Buck LB (2001) A candidate taste receptor gene near a sweet taste locus. Nat Neurosci 4:492–498

Mori K, Nagao H, Yoshihara Y (1999) The olfactory bulb: coding and processing of odor molecule information. Science 286:711–715

Mucignat-Caretta C (2002) Modulation of exploratory behavior in female mice by protein-borne male urinary molecules. J Chem Ecol 28:1853–1863

Mucignat-Caretta C, Caretta A (1999a) Chemical signals in male house mice urine: protein-bound molecules modulate interaction between sexes. Behaviour 136:331–343

Mucignat-Caretta C, Caretta A (1999b) Urinary chemical cues affect light avoidance behaviour in male laboratory mice, Mus musculus. Anim Behav 57:765–769

Mucignat-Caretta C, Caretta A, Cavaggioni A (1995) Acceleration of puberty onset in female mice by male urinary proteins. J Physiol 486:517–522

Mucignat-Caretta C, Caretta A, Baldini E (1998) Protein-bound male urinary pheromones: differential responses according to age and gender. Chem Senses 23:67–70

Mucignat-Caretta C, Cavaggioni A, Caretta A (2004) Male urinary chemosignals differentially affect aggressive behavior in male mice. J Chem Ecol 30:777–791

Nelson G, Hoon MA, Chandrashekar J, Zhang Y, Ryba NJ, Zuker CS (2001) Mammalian sweet taste receptors. Cell 106:381–390

Nelson G, Chandrashekar J, Hoon MA, Feng L, Zhao G, Ryba NJ, Zuker CS (2002) An amino-acid taste receptor. Nature 416:199–202

Niimura Y, Nei M (2003) Evolution of olfactory receptor genes in the human genome. Proc Natl Acad Sci U S A 100:12235–12240

Novotny M, Harvey S, Jemiolo B, Alberts J (1985) Synthetic pheromones that promote inter-male aggression in mice. Proc Natl Acad Sci U S A 82:2059–2061

Novotny M, Jemiolo B, Harvey S, Wiesler D, Marchlewska-Koj A (1986) Adrenal-mediated endogenous metabolites inhibit puberty in female mice. Science 231:722–725

Novotny M, Harvey S, Jemiolo B (1990) Chemistry of male dominance in the house mouse, Mus domesticus. Experientia 46:109–113

Novotny MV, Jemiolo B, Wiesler D, Ma W, Harvey S, Xu F, Xie TM, Carmack M (1999) A unique urinary constituent, 6-hydroxy-6-methyl-3-heptanone, is a pheromone that accelerates puberty in female mice. Chem Biol 6:377–383

Okamoto T, Sekiyama N, Otsu M, Shimada Y, Sato A, Nakanishi S, Jingami H (1998) Expression and purification of the extracellular ligand binding region of metabotropic glutamate receptor subtype 1. J Biol Chem 273:13089–13096

Okere CO, Kaba H (2000) Increased expression of neuronal nitric oxide synthase mRNA in the accessory olfactory bulb during the formation of olfactory recognition memory in mice. Eur J Neurosci 12:4552–4556

Pantages E, Dulac C (2000) A novel family of candidate pheromone receptors in mammals. Neuron 28:835–845

Pearse-Pratt R, Schellinck H, Brown R, Singh PB, Roser B (1998) Soluble MHC antigens and olfactory recognition of genetic individuality: the mechanism. Genetica 104:223–230

Peleg S, Varon D, Ivanina T, Dessauer CW, Dascal N (2002) G(alpha)(i) controls the gating of the G protein-activated K(+) channel, GIRK. Neuron 33:87–99

Perez-Reyes E (2003) Molecular physiology of low-voltage-activated t-type calcium channels. Physiol Rev 83:117–161

Pin JP, Duvoisin R (1995) The metabotropic glutamate receptors: structure and functions. Neuropharmacology 34:1–26

Poran NS, Vandoros A, Halpern M (1993) Nuzzling in the gray short-tailed opossum. I: Delivery of odors to vomeronasal organ. Physiol Behav 53:959–967

Pouille F, Cavelier P, Desplantez T, Beekenkamp H, Craig PJ, Beattie RE, Volsen SG, Bossu JL (2000) Dendro-somatic distribution of calcium-mediated electrogenesis in purkinje cells from rat cerebellar slice cultures. J Physiol 527:265–282

Preti G, Wysocki CJ, Barnhart KT, Sondheimer SJ, Leyden JJ (2003) Male axillary extracts contain pheromones that affect pulsatile secretion of luteinizing hormone and mood in women recipients. Biol Reprod 68:2107–2113

Rasband MN, Shrager P (2000) Ion channel sequestration in central nervous system axons. J Physiol 525:63–73

Ressler KJ, Sullivan SL, Buck LB (1994) Information coding in the olfactory system: evidence for a stereotyped and highly organized epitope map in the olfactory bulb. Cell 79:1245–1255

Richards DB, Stevens DA (1974) Evidence for marking with urine by rats. Behav Biol 12:517–523

Rodriguez I (2004) Pheromone receptors in mammals. Horm Behav 46:219–230

Rodriguez I, Mombaerts P (2002) Novel human vomeronasal receptor-like genes reveal species-specific families. Curr Biol 12:R409–R411

Rodriguez I, Feinstein P, Mombaerts P (1999) Variable patterns of axonal projections of sensory neurons in the mouse vomeronasal system. Cell 97:199–208

Rodriguez I, Greer CA, Mok MY, Mombaerts P (2000) A putative pheromone receptor gene expressed in human olfactory mucosa. Nat Genet 26:18–19

Rodriguez I, Del Punta K, Rothman A, Ishii T, Mombaerts P (2002) Multiple new and isolated families within the mouse superfamily of V1r vomeronasal receptors. Nat Neurosci 5:134–140

Rossler P, Kroner C, Krieger J, Lobel D, Breer H, Boekhoff I (2000) Cyclic adenosine monophosphate signaling in the rat vomeronasal organ: role of an adenylyl cyclase type VI. Chem Senses 25:313–322

Runnenburger K, Breer H, Boekhoff I (2002) Selective G protein beta gamma-subunit compositions mediate phospholipase C activation in the vomeronasal organ. Eur J Cell Biol 81:539–547

Ryba NJ, Tirindelli R (1997) A new multigene family of putative pheromone receptors. Neuron 19:371–379

Saito TR, Moltz H (1986) Copulatory behavior of sexually naive and sexually experienced male rats following removal of the vomeronasal organ. Physiol Behav 37:507–510

Sasaki K, Okamoto K, Inamura K, Tokumitsu Y, Kashiwayanagi M (1999) Inositol-1,4,5-trisphosphate accumulation induced by urinary pheromones in female rat vomeronasal epithelium. Brain Res 823:161–168

Sato T, Shimada Y, Nagasawa N, Nakanishi S, Jingami H (2003) Amino acid mutagenesis of the ligand binding site and the dimer interface of the metabotropic glutamate receptor 1. Identification of crucial residues for setting the activated state. J Biol Chem 278:4314–4321

Savic I, Berglund H, Gulyas B, Roland P (2001) Smelling of odorous sex hormone-like compounds causes sex-differentiated hypothalamic activations in humans. Neuron 31:661–668

Schild D, Restrepo D (1998) Transduction mechanisms in vertebrate olfactory receptor cells. Physiol Rev 78:429–466

Schwende FJ, Wiesler D, Jorgenson JW, Carmack M, Novotny M (1986) Urinary volatile constituents of the house mouse, Mus musculus, and their endocrine dependency. J Chem Ecol 12:277–296

Serizawa S, Ishii T, Nakatani H, Tsuboi A, Nagawa F, Asano M, Sudo K, Sakagami J, Sakano H, Ijiri T, Matsuda Y, Suzuki M, Yamamori T, Iwakura Y (2000) Mutually exclusive expression of odorant receptor transgenes. Nat Neurosci 3:687–693

Serizawa S, Miyamichi K, Nakatani H, Suzuki M, Saito M, Yoshihara Y, Sakano H (2003) Negative feedback regulation ensures the one receptor-one olfactory neuron rule in mouse. Science 302:2088–2094

Shahan K, Gilmartin M, Derman E (1987) Nucleotide sequences of liver, lachrymal, and submaxillary gland mouse major urinary protein mRNAs: mosaic structure and construction of panels of gene-specific synthetic oligonucleotide probes. Mol Cell Biol 7:1938–1946

Shapiro LS, Roland RM, Li CS, Halpern M (1996) Vomeronasal system involvement in response to conspecific odors in adult male opossums, Monodelphis domestica. Behav Brain Res 77:101–113

Silvotti L, Giannini G, Tirindelli R (2005) The vomeronasal receptor V2R2 does not require escort molecules for heterologous expression. Chem Senses (in press)

Simerly RB (1990) Hormonal control of neuropeptide gene expression in sexually dimorphic olfactory pathways. Trends Neurosci 13:104–110

Singer AG, Agosta WC, Clancy AN, Macrides F (1987) The chemistry of vomeronasally detected pheromones: characterization of an aphrodisiac protein. Ann N Y Acad Sci 519:287–298

Singer AG, Beauchamp GK, Yamazaki K (1997) Volatile signals of the major histocompatibility complex in male mouse urine. Proc Natl Acad Sci U S A 94:2210–2214

Speca DJ, Lin DM, Sorensen PW, Isacoff EY, Ngai J, Dittman AH (1999) Functional identification of a goldfish odorant receptor. Neuron 23:487–498

Spehr M, Hatt H, Wetzel CH (2002) Arachidonic acid plays a role in rat vomeronasal signal transduction. J Neurosci 22:8429–8437

Stephan H, Baron G, Frahm HD (1982) Comparison of brain structure volumes in insectivora and primates. II. Accessory olfactory bulb (AOB). J Hirnforsch 23:575–591

Stern K, McClintock MK (1998) Regulation of ovulation by human pheromones. Nature 392:177–179

Stowers L, Holy TE, Meister M, Dulac C, Koentges G (2002) Loss of sex discrimination and male-male aggression in mice deficient for TRP2. Science 295:1493–1500

Swann JM (1997) Gonadal steroids regulate behavioral responses to pheromones by actions on a subdivision of the medial preoptic nucleus. Brain Res 750:189–194

Tirindelli R, Ryba NJ (1996) The G-protein gamma-subunit G gamma 8 is expressed in the developing axons of olfactory and vomeronasal neurons. Eur J Neurosci 8:2388–2398

Tirindelli R, Mucignat-Caretta C, Ryba NJ (1998) Molecular aspects of pheromonal communication via the vomeronasal organ of mammals. Trends Neurosci 21:482–486

Trinh K, Storm DR (2003) Vomeronasal organ detects odorants in absence of signaling through main olfactory epithelium. Nat Neurosci 6:519–525

Trombley PQ, Westbrook GL (1991) Voltage-gated currents in identified rat olfactory receptor neurons. J Neurosci 11:435–444

Trotier D, Doving KB, Ore K, Shalchian-Tabrizi C (1998) Scanning electron microscopy and gramicidin patch clamp recordings of microvillous receptor neurons dissociated from the rat vomeronasal organ. Chem Senses 23:49–57

Tsuji Y, Shimada Y, Takeshita T, Kajimura N, Nomura S, Sekiyama N, Otomo J, Usukura J, Nakanishi S, Jingami H (2000) Cryptic dimer interface and domain organization of the extracellular region of metabotropic glutamate receptor subtype 1. J Biol Chem 275:28144–28151

Van Der Lee S, Boot LM (1955) Spontaneous pseudopregnancy in mice. Acta Physiol Pharmacol Neerl 4:442–444

Vandenbergh JG (1969) Male odor accelerates female sexual maturation in mice. Endocrinology 84:658–660

Vassalli A, Rothman A, Feinstein P, Zapotocky M, Mombaerts P (2002) Minigenes impart odorant receptor-specific axon guidance in the olfactory bulb. Neuron 35:681–696

Vassar R, Chao SK, Sitcheran R, Nunez JM, Vosshall LB, Axel R (1994) Topographic organization of sensory projections to the olfactory bulb. Cell 79:981–991

Verberne G (1976) Chemocommunication among domestic cats, mediated by the olfactory and vomeronasal senses. II. The relation between the function of Jacobson's organ (vomeronasal organ) and Flehmen behaviour. Z Tierpsychol 42:113–128

von Campenhausen H, Mori K (2000) Convergence of segregated pheromonal pathways from the accessory olfactory bulb to the cortex in the mouse. Eur J Neurosci 12:33–46

Wang F, Nemes A, Mendelsohn M, Axel R (1998) Odorant receptors govern the formation of a precise topographic map. Cell 93:47–60

Waterston RH, Lindblad-Toh K, Birney E, Rogers J, Abril JF, et al (2002) Initial sequencing and comparative analysis of the mouse genome. Nature 420:520–562

Wekesa KS, Anholt RR (1997) Pheromone regulated production of inositol-(1, 4, 5)-trisphosphate in the mammalian vomeronasal organ. Endocrinology 138:3497–3504

Wekesa KS, Anholt RR (1999) Differential expression of G proteins in the mouse olfactory system. Brain Res 837:117–126

Wekesa KS, Miller S, Napier A (2003) Involvement of G(q/11) in signal transduction in the mammalian vomeronasal organ. J Exp Biol 206:827–832

Westberry JM, Meredith M (2003) Pre-exposure to female chemosignals or intracerebral GnRH restores mating behavior in naive male hamsters with vomeronasal organ lesions. Chem Senses 28:191–196

White JH, Wise A, Main MJ, Green A, Fraser NJ, Disney GH, Barnes AA, Emson P, Foord SM, Marshall FH (1998) Heterodimerization is required for the formation of a functional GABA(B) receptor. Nature 396:679–682

Whitten WK (1956) Modification of the oestrous cycle of the mouse by external stimuli associated with the male. J Endocrinol 13:399–404

Winberg J, Porter RH (1998) Olfaction and human neonatal behaviour: clinical implications. Acta Paediatr 87:6–10

Wirsig-Wiechmann CR, Wiechmann AF, Eisthen HL (2002) What defines the nervus terminalis? Neurochemical, developmental, and anatomical criteria. Prog Brain Res 141:45–58

Wu Y, Tirindelli R, Ryba NJ (1996) Evidence for different chemosensory signal transduction pathways in olfactory and vomeronasal neurons. Biochem Biophys Res Commun 220:900–904

Wysocki CJ, Lepri JJ (1991) Consequences of removing the vomeronasal organ. J Steroid Biochem Mol Biol 39:661–669

Wysocki CJ, Wellington JL, Beauchamp GK (1980) Access of urinary nonvolatiles to the mammalian vomeronasal organ. Science 207:781–783

Wysocki CJ, Nyby J, Whitney G, Beauchamp GK, Katz Y (1982) The vomeronasal organ: primary role in mouse chemosensory gender recognition. Physiol Behav 29:315–327

Yamazaki K, Boyse EA, Mike V, Thaler HT, Mathieson BJ, Abbott J, Boyse J, Zayas ZA, Thomas L (1976) Control of mating preferences in mice by genes in the major histocompatibility complex. J Exp Med 144:1324–1335

Young JM, Shykind BM, Lane RP, Tonnes-Priddy L, Ross JA, Walker M, Williams EM, Trask BJ (2003) Odorant receptor expressed sequence tags demonstrate olfactory expression of over 400 genes, extensive alternate splicing and unequal expression levels. Genome Biol 4:R71

Zhang X, Firestein S (2002) The olfactory receptor gene superfamily of the mouse. Nat Neurosci 5:124–133

Zhang Y, Hoon MA, Chandrashekar J, Mueller KL, Cook B, Wu D, Zuker CS, Ryba NJ (2003) Coding of sweet, bitter, and umami tastes: different receptor cells sharing similar signaling pathways. Cell 112:293–301

Zhao GQ, Zhang Y, Hoon MA, Chandrashekar J, Erlenbach I, Ryba NJ, Zuker CS (2003) The receptors for mammalian sweet and umami taste. Cell 115:255–266

Zhou A, Moss RL (1997) Effect of urine-derived compounds on cAMP accumulation in mouse vomeronasal cells. Neuroreport 8:2173–2177

Zou DJ, Feinstein P, Rivers AL, Mathews GA, Kim A, Greer CA, Mombaerts P, Firestein S (2004) Postnatal refinement of peripheral olfactory projections. Science 304:1976–1979

Zou Z, Horowitz LF, Montmayeur JP, Snapper S, Buck LB (2001) Genetic tracing reveals a stereotyped sensory map in the olfactory cortex. Nature 414:173–179

Rev Physiol Biochem Pharmacol (2005) 154:37–72
DOI 10.1007/s10254-005-0041-0

Wolfgang Meyerhof

Elucidation of mammalian bitter taste

Published online: 20 July 2005
© Springer-Verlag 2005

Abstract A family of approximately 30 TAS2R bitter taste receptors has been identified in mammals. Their genes evolved through adaptive diversification and are linked to chromosomal loci known to influence bitter taste in mice and humans. The agonists for various TAS2Rs have been identified and all of them were established as bitter tastants. TAS2Rs are broadly tuned to detect multiple bitter substances, explaining, in part, how mammals can recognize numerous bitter compounds with a limited set of receptors. The TAS2Rs are expressed in a subset of taste receptor cells, which are distinct from those mediating responses to other taste qualities. However, cells devoted to the detection of sweet, umami, and bitter stimuli share common signal transduction components. Transgenic expression of a human TAS2R in sweet or bitter taste receptor-expressing cells of mice induced either strong attraction or aversion to the receptor's cognate bitter tastant. Thus, dedicated taste receptor cells appear to function as broadly tuned detectors for bitter substances and are wired to elicit aversive behavior.

Introduction

The sense of taste allows mammals to evaluate their food; they must discriminate nutrients from toxins. Taste in humans has been categorized into the five distinct basic taste qualities or modalities: sweet, sour, salty, bitter, and recently umami (the taste of monosodium glutamate) was added (Lindemann 1996). The existence of additional basic tastes is still debated (Lindemann 1996; Schiffman 2000). It is assumed that each of the basic tastes serves distinct functions. Umami and sweet tastes are said to be caloric detectors, sensing protein-rich or carbohydrate-rich food. These tastes elicit a positive hedonic tone, activate the brain's reward and reinforcing systems, and induce feeding behavior (Saper et al. 2002). Salt taste is one of the powerful mechanisms that mammals developed during evolution away from the

Wolfgang Meyerhof (✉)
German Institute of Human Nutrition Potsdam-Rehbrücke, Department of Molecular Genetics, Arthur-Scheunert-Allee 114–116, 14558 Nuthetal, Germany
e-mail: meyerhof@mail.dife.de · Tel.: +49-33200-88282 · Fax: +49-33200-88384

sea to land-life to maintain their sodium levels and is particularly important in herbivores (Denton 1982). It serves as a detector of food sources containing sodium and other mineral ions. Sour taste detects protons and contributes to the recognition of food, but becomes increasingly unpleasant at high concentrations (Lindemann 1996). Its physiological role is presumably the recognition of unripe or spoiled food. Bitter taste is assumed to detect toxins in food; in fact, many poisonous substances elicit bitter taste. Most of the bitter compounds arise from plants. Some 10% of plants may contain toxic glycosides or alkaloids (Kingsbury 1964). Alone 2500 plant species synthesize cyanogenic glycosides (Zagrobelny et al. 2004). Bitter toxins evolved in plants as chemical defense systems against herbivores and pathogens (Biere et al. 2004). However, bitter tasting compounds not only originate from plants, they can also occur in animals (Murata and Sata 2000). Certain insects can synthesize cyanogenic glycosides de novo or sequester it from plants and use these compounds in their own defense against predators (Zagrobelny et al. 2004).

Strong bitter taste is highly aversive, while mild bitter taste in certain human food items or beverages may be enjoyed. Bitter taste is innate; human neonates react with stereotypic rejection responses indicative of a strong negative hedonic tone (Steiner 1994). All animals react with aversive or repulsive behavioral patterns to compounds that humans categorize as bitter, although some herbivores are quite insensitive (Nolte et al. 1994; Lindemann 1996). Bitter avoidance behavior is already evident in invertebrates. Many water-soluble substances that are toxic and taste bitter to humans are repulsive to the worm *Caenorhabditis elegans* and to the fly *Drosophila melenogaster* (Matsunami and Amrein 2003; Hilliard et al. 2004).

How the detection of these bitter compounds by the gustatory system is accomplished and how gustatory information is transmitted to and perceived by the central nervous system is not well understood. However, the recent cloning of mammalian bitter taste receptors, particularly through the pioneering work of Nicholas Ryba (Bethesda) and Charles Zuker (San Diego) allowed researchers to develop a number of experimental tools. Their use has enabled rapid progress in taste research over the last couple of years. The objective of this review is to summarize the current knowledge about the bitter taste receptors, their phylogeny, structure–function relationship, site of expression, signal transduction, and the implications that these receptors have for the neural code of bitter taste. These new findings will be discussed in view of existing knowledge about bitter taste physiology.

Taste anatomy

Taste anatomy has been reviewed in detail (Miller 1995); here only a brief outline will be given. The primary structures of taste reception are the taste buds, which are found in all vertebrates except hagfishes (Northcutt 2004). Taste buds are assemblies of approximately 100 cells, which, in humans and other mammals, are present on the tongue, the soft palate, epiglottis, pharynx, and larynx. Taste buds are embedded in the epithelium such that their basolateral part is shielded from the oral cavity by tight junctions. At least four different cell types have been identified by morphological criteria, called type I to type IV cells. However, cells of a given type appear not to be uniform and may fall into functionally distinct subclasses (Clapp et al. 2004). Type IV cells are considered to be progenitor cells placed at the base of the taste bud. The other types of taste bud cells contact the oral cavity only with their apical microvilli containing tips in a depression called a taste pore. Only in the taste pore do interactions of tastants with taste receptor cells take place. Therefore, type I to type III cells are candidate taste receptor cells (TRCs). Recently, various signal transduction molecules have been detected in most type II cells and a small subset of type III cells suggesting that the TRCs for umami, sweet, and bitter taste are among these cells (Yang et al. 1999; Clapp et al. 2004). TRCs are not neurons but specialized epithelial cells, and are accordingly designated as secondary sensory cells. In addition to their organization in taste buds, TRCs may also exist as solitary chemosensory cells in the developing mammalian gustatory epithelium (Sbarbati et al. 1999; Sbarbati and Osculati 2003) and chemosensory clusters in the larynx (Sbarbati et al. 2004).

On the tongue, taste buds are organized in specialized folds and protrusions, the gustatory papillae (Miller 1995). The three types of chemosensory papillae are the fungiform, foliate, and vallate papillae. In humans the latter are also called circumvallate papillae, because they are completely surrounded by moats. Circumvallate papillae contain approximately 250 taste buds and are present in a V-shaped line across the root of the tongue. Their numbers vary greatly in humans from 3 to 18; rats possess only one vallate papilla centered at the root of the tongue. The foliate papillae located at both posterior margins of the tongue consist of 2 to 9 alternating ridges and fissures in humans and contain approximately 120 taste buds per ridge. The number of buds per ridge differs considerably among subjects. Ducts of the intralingual von Ebner's glands empty into the troughs of vallate and foliate papillae. The secretion constitutes a perireceptor milieu in which tastants dissolve and are transported to or from the receptor sites. Thus, the secretion maintains taste sensitivity and may influence gustatory sensation (Schmale et al. 1990 #117; Matsuo 2000). Fungiform papillae appear as elevations of about 0.5 mm in diameter spread unevenly about the tongue surface, such that the tip of the tongue possesses a higher density of fungiform papillae than the midposterior region. The number of taste buds per fungiform papillae differs between individuals. Depending on the location along the anterior/posterior axis average counts range from 3.4 to 2.6 buds per fungiform papilla (Miller 1995). Based on the different locations of the various papillae attempts have been made to construct chemotopic maps of the tongue as they are seen in various textbooks. They should be considered with caution since threshold determinations in subjects revealed very little variation in the sensitivities for salty and sweet stimuli across the tongue. The sensitivity for quinine was only less than 2-fold elevated at the base of the tongue compared with the tip, while sour taste was about 2-fold less sensitive at the base compared with the margins of the tongue (von Skramlik 1926).

The internal organization of taste buds raises many questions. As secondary sensory cells, TRCs connect to afferent sensory nerve fibers via chemical synapses, although the neurotransmitter is not firmly established (Roper 1992). Serotonin has been identified as a transmitter in mouse taste buds (Huang et al. 2005). However, it may serve intercellular communication within the bud and not excite the afferents. Mammalian TRCs turn over fast, their lifespan being about 10 days (Lindemann 2001). Consequently, the termini of the afferent nerves have to detach from aged cells, and find and connect to developing cells. These processes raise concern about the stimulus specificity of innervation. Are definite fibers dedicated to TRCs responding to stimuli of only one taste quality? What developmental program do TRCs and afferent nerves execute? In light of recordings that showed that many TRCs and fibers are generalists, responding to stimuli of more than one taste quality (Scott and Giza 2000), it may be asked to what extent specificity is required. If so, do the nerve fibers attach to TRCs of the same specificity as before, and what molecules serve as markers? If the receptor molecules themselves represent such markers as in the olfactory system (Mombaerts 2004), they are required to be present not only at the apical but also at the basolateral part of the TRCs. Immunocytochemical localization of the epithelial sodium channel, ENaC, the putative "receptor" for salty taste (Kretz et al. 1999), and of T1R3 (Max et al. 2001), a subunit of the sweet and umami taste receptor, is consistent with this assumption. Future functional approaches elucidating taste bud development, the connectivities between TRCs and the afferent fibers, and the rules guiding them will be essential to unravel intercellular communication within the buds and the code of gustatory information transmission.

Branches of the three cranial nerves innervate the taste buds, such that individual axons supply several TRCs, which may be located in different taste buds (Martin 1989; Smith and Frank 1993; Miller 1995; Whitehead et al. 1999). The chorda tympani of the seventh cranial or facial nerve originating from the geniculate ganglion innervates the anterior two-thirds of the tongue and may transmit predominantly chemosensory information from the fungiform papillae. The greater superficial petrosal branch of this nerve innervates the taste buds of the palate. The posterior one-third of the tongue receives supply from the ninth, the glossopharyngeal nerve, which may carry mostly gustatory information from the vallate and foliate papillae. The cell bodies of this nerve are located in the petrosal ganglion. Vagal axons from neurons in the nodose ganglion innervate taste buds in the epiglottis, pharynx, and larynx. The first order pseudounipolar ganglionic neurons make contact with second order neurons in the rostral part of nucleus tractus solitarius (NTS), in a region referred to as the gustatory nucleus, located in the medulla. In humans, the second order neurons project monosynaptically on third order neurons in the parvocellular part of the ventral posteromedial thalamic nucleus (VPMpc) on the ipsilateral side. This is a remarkable difference to the taste system in rodents, which has an obligatory relay from the NTS to the pontine parabrachial taste nucleus, in turn projecting to the VPMpc (Rolls 1995). Here, the gustatory information is relayed to the primary gustatory cortex in the anterior insula and the frontal operculum. A secondary cortical taste area located in the caudolateral orbitofrontal cortex receives inputs from the primary taste area (Rolls 1995). The activities of subsets of neurons at all of the aforementioned levels reflect primary sensation, transmission, coding, processing, and perception of gustatory information. However, how this is achieved is far from being clear. Peripheral and even more central neurons appear to be broadly tuned to stimuli of different qualities, but may respond best to only one (Smith and St John 1999), which has led to opposite theories about taste quality coding. The labeled line theory states that neurons of distinct taste qualities act independently from each other (see below), whereas the across neuron pattern theory assumes that all neurons contribute to the quality of a taste irrespective of whether they are excited or not. Nonetheless, in primates the identity and intensity of a taste is likely to be represented by neuronal activities in the primary taste cortex (Scott et al. 1999), while the secondary taste area has been suggested to reflect reward-related aspects of taste sensation (Rolls 2004). Using functional magnetic resonance tomography the representation of the five basic taste qualities have been analyzed in humans (Schoenfeld et al. 2004). The authors observed intense inter-individual differences, but stable image patterns over time within subjects. Nonetheless, they were able to visualize different representations with some considerable overlap for all primary tastes, i.e. salty, sour, sweet, bitter, and umami within the primary taste cortex. Interestingly, the region activated more by the aversive bitter stimulus than by any other taste stimulus appeared to lie most laterally of the primary taste cortex and most far apart from the region activated by the attractive sweet stimulus. However, in light of limitations introduced by the extreme folding of this part of the cortex, movement of subjects during taste stimulus presentation, and inter-subject variability, they were careful to emphasize that their findings do not firmly support a chemotopical organization of the primary taste cortex.

Bitter substances

Numerous chemically diverse compounds elicit bitter taste. Unfortunately, to the best of the author's knowledge, no complete database exists that has archived the known bitter tasting chemicals together with their molecular and physiological properties. Also, a lack of (quantitative) structure–activity studies of bitter compounds in general limits our knowledge about the structural parameters that make chemicals bitter. In an early review, von Skramlik already identified $-NO_2$, $N\equiv$, $=N\equiv$, $-SH$, $-S-$, $-S-S-$, $=CS$, and the cyclic $CH_2-CO-O-N$ group as amarogenic (von Skramlik 1926). However, he was cautious to emphasize that the presence of such a group alone may not predict bitterness. Other parameters apply as well, including hydrophobicity, stereoisomeric properties, repetitive presence of an amarogenic group or combined presence of different amarogenic groups in a chemical (von Skramlik 1926). Based on the structural similarity of some sweet to bitter tasting compounds an AH-B model for "the bitter receptor" has been proposed (Tancredi et al. 1979) very similar to that of the sweet taste receptor (Shallenberger and Acree 1967). In this model, bitter tasting molecules should contain an entity with a hydrogen donor site, called A-H, and a hydrogen acceptor site, called B, spaced about 0.3 nm apart (Tancredi et al. 1979). Although several structurally different bitter tasting molecules have been fitted into the receptor site, including 2,3-naphthosaccharin, 6-Br-saccharin, quinine, and glucopyranosides (Tancredi et al. 1979; Matsumoto et al. 1986) its general applicability remains to be seen. The great number of bitter compounds along with their structural diversity, varying fundamental parameters such as size, charge, and hydrophobicity, clearly suggested the existence of a family of receptors, instead of only one receptor recognizing most or even all bitter tasting compounds (Adler et al. 2000; Matsunami et al. 2000). A compilation of various bitter compounds is depicted in Table 1. However, it should be stressed here that many more compounds than those listed taste bitter, including terpenes, diterpenes, triterpenes, sesquiterpene lactones, flavones, bitter acids of the humulone series, and glycosides, among them amarogentin, the most intense bitter compound known (Belitz and Wieser 1985).

Many bitter tasting compounds are contained in human foodstuffs. Obviously, some of them are present, because they are of plant origin (vegetables, cabbage, green salads, beer, herbal extracts in alcoholic drinks; Gienapp and Schröder 1975; Rouseff 1980; Belitz and Wieser 1985; Lechtenberg and Nahrstedt 1999; Gutierrez-Rosales et al. 2003; Serafini et al. 2003; Moriyama and Oba 2004; Lesschaeve and Noble 2005); others are generated during food processing including heating (roasting of coffee, Maillard reaction products of amino acids and sugars; Frank and Hofmann 2002; Schieberle and Hofmann 2003) and fermentation (cheese; Habibi-Najafi and Lee 1996; Engel et al. 2001). Still others appear during the aging of food (rancid fat through lipolysis or lipid oxidation, bitterness production as part of the plant defense system during storage, peptides from hydrolyzed proteins; Belitz and Wieser 1985; Drewnowski and Gomez-Carneros 2000; Czepa and Hofmann 2003, 2004). Since bitter taste is generally linked to aversive responses, elevated levels of bitter compounds may lead to rejection of certain food items by the consumers (Rozin and Vollmecke 1986), thereby protecting them from ingesting potentially toxic compounds such as alkaloids or cyanogenic glycosides or spoiled food. But as usual, there are two sides of the coin. Quite a number of bitter tasting compounds have recently been compiled (Barratt-Fornell and Drewnowski 2002) that possibly exert health benefits (Table 2). The authors pointed out that the role of these plant-based chemical substances or phytonutrients is to protect the plant from stress of various kinds, fungal or bacterial infections, or predators. Apparently, the right dose will heal or harm. At high concentrations, phytonutrients can also exert undesirable actions (Table 2). Nonetheless, the food industry employs various debittering

Rev Physiol Biochem Pharmacol (2005) 154:37–72

Table 1 Selected bitter compounds and their human threshold values

Compound class	Substance	Recognition threshold (mM)
Hydroxy fatty acids, fatty acids	9-Hydroxy-10t,12c-octadecadienoic acid	6.5–8.0
	9,10,13-Trihydroxyoctadeca-10-enoic acid	0.6–0.9
	9c-Octadecenoic acid	9.0–12.0
	9c,12c,15c-Octadecatrienoic acid	0.6–1.2
	5c,8c,11c,14c-Eicosatetraenoic acid	6.0–8.0
Peptides	Ala-Leu	18–22
	Leu-Val-Leu	2.0
	Phe-Gly-Phe-Gly	1.0–1.5
Amino acids	L-valine	~20
	L-phenylalanine	4.5–7.0
	L-tyrosine	4.0–6.0
	L-tyrosine	4.0–6.0
Amines	Propylamine	15–25
	Butylamine	4.0–8.0
	Pentylamine	1.5–2.0
	Dodecanylamine	0.3–0.5
	Diacetylamine	5.0–15
	Triacetylamine	2.0–3.0
Azacycloalkanes	Azacyclopentane	8.0–12
	Azacyclohexane	0.6–1.0
N-Heterocyclic compounds	Pyrazole	15–20
	Imidazole	4.0–8.0
	Piperidine	8.0–12
	Piperazine	20–30
	Pyridine	1.0–3.0
	Adenine	2.0–4.0
	Adenosine	3.0–6.0
	Hypoxanthine	4.5–6.0
Amides	Propionamide	50–55
	N-methylacetamide	10–15
	Benzamide	0.8–1.0
	Denatonium benzoate	0.00001–0.00002
	Thioacetamide	1.0–2.0
	N-phenylthioacetamide	0.01–0.02
	Delta-valerolactam-2-piperidinone	3.0–4.0
	Saccharin	~0.3[a]
Ureas, thioureas, carbamides	Urea	60–70
	N-methylurea	35–40
	N-ethylurea	20–25
	N-butylurea	5.0–7.5
	N-phenylurea	4.0–6.0
	N-methylthiourea	0.5–0.6
	N,N-dimethylcarbamate	5.0–8.0
Esters, lactones	Ethylbenzoate	2.0–5.0
	4-Ethylhydroxybenzoate	4.0–6.0
	4-Aminohydroxybenzoate	2.0–5.0
	Butyrolactone	50–60
	Ethyl-γ-butyrolactone	10
	Butyl-γ-butyrolactone	0.2–2.0
Carbonyl compounds	2-Pentanone	25–35
	2-Octanone	3.5–4.0
	2-Decanone	1.0–1.5
	Cyclooctanone	2.0–4.0

Table 1 (continued)

Compound class	Substance	Recognition threshold (mM)
Phenols	1,2-Dihydroxybenzene	5.0–10
	1,3-Dihydroxybenzene	5.0–15
	1,2,3-Trihydroxybenzene	6.0–8.0
Crown ethers	12-Crown-4-ether	12–15
	18-Crown-6-ether	0.5–0.7
	Dibenzo-18-crown-6-ether	0.01–0.013
Alkaloids	Caffeine	0.2
	Nicotine	0.019
	Atropine	0.1
	Cocaine	0.5
	Quinine	0.01
	Strychnine	0.002
	Morphine	0.5
	Cholcicine	0.01
Metal ions	KI	6.4
	$CaSO_4$	3.7
	$MgSO_4$	4.2

[a]Our own unpublished results
The data were compiled from Belitz and Wieser (1985) and von Skramlik (1926)

processes, including breeding techniques to avoid and enzymatic methods to remove bitterness from foodstuffs (Drewnowski and Gomez-Carneros 2000; Raksakulthai and Haard 2003).

Humans do not solely ingest bitter compounds with their food, but also through their medications. Particularly children and chronic patients suffer from the bitterness of their medications and have reduced quality of life. Since compliance with the medication determines its therapeutic success, the pharmaceutical industry developed strategies to physically mask the medicines' bitter taste, including microsphere formation and coating (Suzuki et al. 2003). Nonetheless, good examples for orally administered bitter tasting drugs are antibiotics such as clarythromycin and erythromycin to treat bacterial infections (Uchida et al. 2003) and protease inhibitors such as ritonavir, indinavir, and saquinavir to treat HIV patients (Schiffman et al. 1999). Thus, alternative methods for alleviating the patient's taste complaints or for increasing the acceptability of bitter phytonutrients in functional food are desired. Specific bitter masking agents would be an alternative; attempts at finding such compounds based on biochemical assays are being made by biotech companies (McGregor 2004). However, these attempts require knowledge about the bitter taste receptors.

Identification of TAS2R bitter taste receptor genes

Common knowledge about bitter taste shaped the search strategies of two groups who finally succeeded in cloning the long sought-for genes encoding putative bitter taste receptors (Adler et al. 2000; Matsunami et al. 2000). Firstly, based on the observation that bitter taste transduction is mediated by the G protein α-gustducin (McLaughlin et al. 1992; Wong et al. 1996a, b; Ming et al. 1998, 1999; Ruiz-Avila et al. 2000), bitter taste receptors must belong to the GPCR family and expressed in α-gustducin-positive TRCs. Secondly, in light of the numerous bitter tastants, not a single gene but rather a distinct gene family was expected to

Table 2 Compilation of various bitter phytonutrients together with their presumed beneficial and detrimental effects on health

Substances	Food source	Possible health benefits	Possible detrimental effects
Carotenoids, including beta-carotene, lycopene, lutein, xanthophylls	Tomato, carrot, sweet potato, watermelon, spinach	Quench singlet oxygen, increase cell–cell communication, decrease risk macular degeneration	
Limonoids, including, d-limonin, limonin glycoside	Citrus fruit	Promote protective enzymes, antiseptic	
Phenol flavonoids, including catechins, anthocyanins, proanthocyanidins, naringin, quercetin, tangeritin, gingko	Tea, berries, wine, citrus fruit, apple, endive, cranberry, onion, kale	Antioxidants, alter tyrosine kinase activity, decrease capillary fragility and permeability, anti-inflammatory	Interference with drug metabolism through inhibition of cytochrome P-450
Isoflavones, including, genistein, diadzein, glycyrrhizin, isocoumarin	Miso, soy milk, soy nut, tofu, licorice, carrot	Metabolize to estrogen-like compounds, decrease tyrosine kinase activity, lower cholesterol	Goitrogenic activity, estrogenic activity
Phenolic acids, including, gallic acid, ellagic acid, caffeic acid, isocoumarin	Tea, strawberry, blueberry, apple, orange, grapefruit, white potato, grape juice, coffee, prune	Increase phase II enzyme activity, antioxidant, inhibit N-nitrosation reactions	
Glucosinoglates/isothiocyanates, including allyl isothiocyanid	Cabbage, kale, brussel sprout, broccoli	Increase phase II enzyme activity, decrease deoxyribonucleic acid (DNA) methylation	Goitrogenic activity, contact dermatitis, carcinogenic activity
Polyphenols, including, tannins and resveratrol	Wine	Inhibit tumor initiation, promotion, and progression, antiinflammatory, antioxidant	Interference with protein absorption, reduction of iron availability

Adapted from (Barratt-Fornell and Drewnowski 2002)

encode bitter taste receptors. Thirdly, genetic mapping of inherited bitter taste differences identified distinct loci determining the sensitivity to sucrose octaacetate, cycloheximide, raffinose undecaacetate, and quinine on mouse chromosome 6 (Lush 1981, 1984, 1986; Lush and Holland 1988) as well as to propylthiouracil (PROP) on human chromosomes 5 and 7 (Reed et al. 1999). Searching the human genome database in intervals linked to bitter taste or in intervals syntenic to the mouse bitter taste loci the first genes encoding members of a new family of GPCRs were identified and called T2Rs or TRBs (Adler et al. 2000; Matsunami et al. 2000). With the available sequence information of these genes several groups cloned additional members of this family from rat, mouse, human, bonobo, chimpanzee, gorilla, orangutan, rhesus macaque, and baboon (Bufe et al. 2002; Conte et al. 2002, 2003; Shi et al. 2003; Parry et al. 2004; Fischer et al. 2005; Go et al. 2005). As many groups more or less simultaneously published their data confusion arose due to the non-consistent or overlapping nomenclature of the bitter taste receptor genes. In this review, the author follows

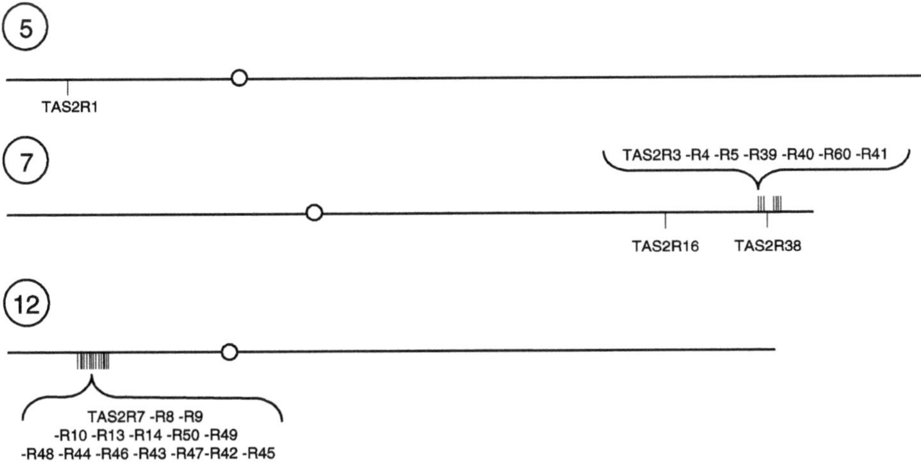

Fig. 1 Chromosomal locations and orientations of the human *TAS2R* genes. Locations of *TAS2R* genes are roughly drawn to scale. Genes in plus orientation are shown above the lines, genes in minus orientation below the lines. *Circles* represent centromeric regions. *Numbers* identify chromosomes. The order of the*TAS2R* genes within the loci is displayed

the T2R nomenclature of Adler et al. (2000), when referring to the rodent genes, and to the TAS2R nomenclature of Bufe et al. (2002), when referring to the human genes. The TAS2R nomenclature will also be used when generally referring to members of this gene family. For a compilation of the different nomenclatures, the readers' attention is drawn to the review of Andres-Barquin and Conte (2004).

Based on latest counts the gene repertoire in humans comprises 25 full-length genes and 11 pseudogenes (Go et al. 2005). With the exception of *TAS2R1* on chromosomes 5, 9 and 15 *TAS2R* genes are present in extended clusters on chromosomes 7 and 12, respectively (Fig. 1; Adler et al. 2000; Matsunami et al. 2000). The mouse genome contains slightly more *T2R* genes, namely 35 genes and six pseudogenes (Go et al. 2005) located on chromosomes 2, 6, and 15 (Adler et al. 2000; Matsunami et al. 2000; Conte et al. 2003). All but two genes, which are found on chromosomes 2 and 15 respectively, are located in two clusters on chromosome 6 (Andres-Barquin and Conte 2004). *TAS2R* genes that lack introns in their coding regions specify proteins composed of approximately 290–330 amino acids. TAS2Rs are structurally diverse; the paralogs display approximately 17–90% sequence identity at the amino acid level, suggesting that different family members may recognize chemicals with widely different structures (Matsunami et al. 2000). However, they share the heptahelical structure and other sequence motifs with one another, classifying them as members of the same GPCR receptor subfamily. Figure 2 shows that, in general, the cytoplasmic part of the putative transmembrane (TM) segments and the intracellular loops are comparably well conserved, consistent with the idea that intracellular coupling to signal transduction proteins is of limited variability. The 20 amino acids present in all or almost all of the TAS2Rs are, depending on their position in the intracellular domains or lower part of the TM segments, likely to be involved in G protein coupling, receptor activation, and folding of the three-dimensional structure of the receptor. The extracellular loops and upper parts of the TM segments are less conserved. This is not surprising, since these parts of the TAS2Rs are likely to form heterogeneous binding motifs for the numerous and structurally diverse bitter compounds. In these regions there are only two highly conserved residues in extracellular loop (EL)-2 forming an N-linked glycosylation site. Thus, unlike most other GPCRs,

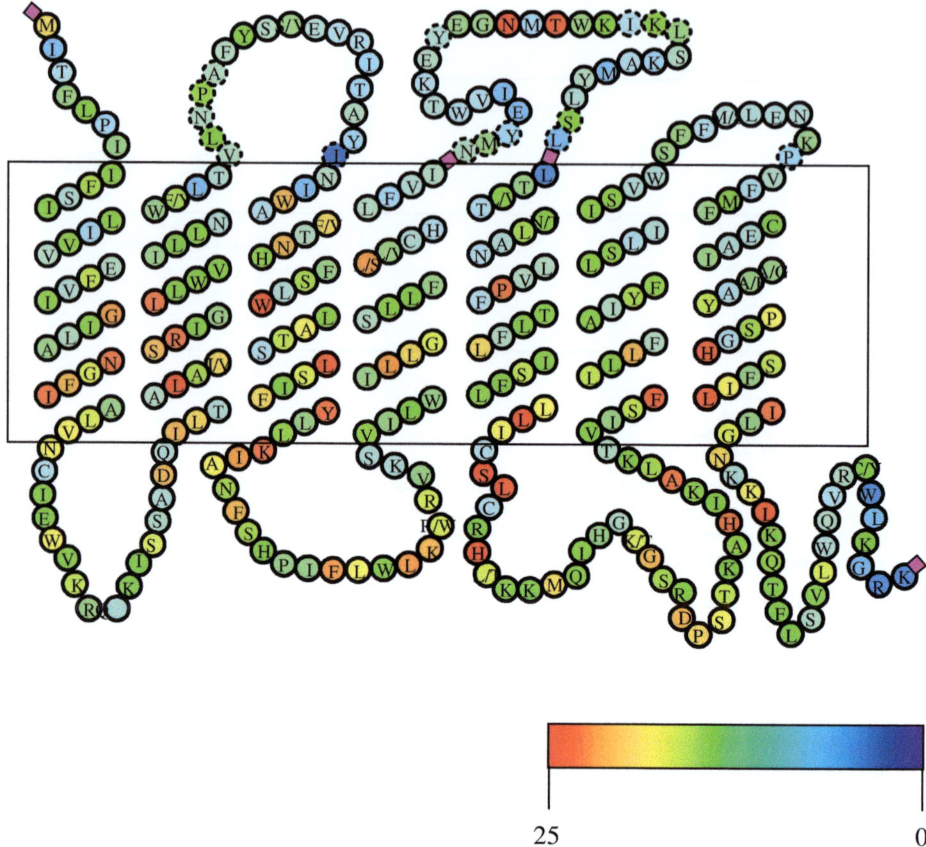

Fig. 2 Schematic representation of hTAS2Rs. Amino acid residues are indicated by *colored circles*. Sequence identity of hTAS2Rs has been analyzed using the pileup program (Genetics Computer group) and is depicted by the color code. *Letters within the circles* indicate the most frequent amino acids (in one letter code) present in the respective positions of the 25 hTAS2Rs. Transmembrane domains were predicted using the TMHMM program http://www.cbs.dtu.dk/services/TMHMM-2.0/). *Dotted circles*, corresponding residues are not present in all hTAS2Rs. *Magenta squares*, regions that contain additional amino acids in a few hTAS2Rs

TAS2Rs do not possess N-linked glycosylation sites in their amino-terminal parts. As in other TM or secretory proteins N-linked glycosylation may be involved in protein function, biosynthesis, and/or protein quality control. The carboxyl-terminal part of TAS2Rs is also highly variable, reflecting possible differences between TAS2Rs in receptor regulation and trafficking.

Phylogeny of TAS2Rs

Sequence comparisons of the human *TAS2R* genes revealed that genes present on the same chromosome are more closely related to one another than to genes on different chromosomes and the presence of a "subfamily" of eight *TAS2R* genes showing particularly high sequence conservation (Bufe et al. 2002). Phylogenetically closer *TAS2R* genes are also closer in their chromosomal location in mice (Shi et al. 2003), supporting the idea that tandem gene duplications account for the generation of new TAS2Rs. Comparisons of the repertoires of the human and mouse *TAS2R* genes revealed that they could be classified into a few main groups (Conte et al. 2002). Whereas for some genes of the groups a one-to-one gene orthology has been observed, the genes of one group showed patterns of species or lineage-specific duplications (Shi et al. 2003). Thus, 11 *mT2Rs* genes cluster with the *hTAS2R14* gene. Vice versa, *hTAS2R43*, *hTAS2R44*, *hTAS2R45*, *hTAS2R46*, *hTAS2R47*, *hTAS2R48*, *hTAS2R49*, and *hTAS2R50* genes cluster together with the *mT2R47* and *mT2R49* genes. Based on this observation it has been argued that genes of the classes showing the one-to-one orthology are possibly necessary for recognizing bitter compounds common to both mice and humans, whereas the genes in the diversity class mediate detection of species-specific bitter substances (Shi et al. 2003). It remains to be seen whether this will turn out to be true for the majority of genes. However, at present there is evidence for and against this conclusion. The properties of the *hTAS2R16* and *hTAS2R38* genes do not support this assumption. These genes belong to a group with a one-to-one orthology and encode receptors for β-glucopyranosides and phenylthiocarbamide, compounds perceived as bitter in humans, while mice are indifferent to them (Mueller et al. 2005). On the other hand the *mT2R5* gene falls, with four other mouse *T2R* genes and the human *TAS2R10* gene, into the diversity group. While the mouse mT2R5 and also its rat ortholog rT2R9 recognize cycloheximide, which is highly aversive in mice, but only moderately bitter to humans (Chandrashekar et al. 2000; Bufe et al. 2002), the human TAS2R10 responds to strychnine, which is bitter to humans, but not particularly aversive to rodents, making it a useful poison for rats and mice.

When the TAS2R genes of other primates were cloned and their sequences analyzed (Parry et al. 2004; Wang et al. 2004; Fischer et al. 2005; Go et al. 2005), several points became clear. Firstly, compared with rodents, primate *TAS2R* genes experienced less selective constraint. Secondly, compared with other primates no significant loss in the number of functional *TAS2R* genes has been observed in humans. This is in contrast to the odorant receptor gene family and has been explained by reduced sensory needs specific for the olfactory system (Fischer et al. 2005). Thirdly, hallmarks of neutral evolution have been observed, including similar rates of synonymous and non-synonymous substitutions among rare polymorphisms, common polymorphisms and substitutions, segregation of pseudogene alleles within species and fixation of loss-of-function mutations (Wang et al. 2004; Fischer et al. 2005; Go et al. 2005). An interesting question is whether the observed ratio of non-synonymous to synonymous substitutions reflects little selective constraint on bitter taste receptors in general, or if they reflect positive selection acting on the part of the protein. It has been concluded by two groups based on the different ratios of TM domains versus the EL domains that in addition to low levels of functional constraint, positive selection may also play a role in the evolution of *TAS2R* genes (Fischer et al. 2005; Go et al. 2005). Positive selection has also been suggested to account for the rapid evolution of the identified *hTAS2R38* nontaster allele, which was then maintained by balancing selection. Although the mechanisms through which the divergent hTAS2R38 alleles were maintained in the human population remained unclear, one possibility was that taster/non-taster heterozygotes gained a fitness advantage through avoiding a larger number of bitter toxins than homozygotes (Wooding et al. 2004).

It has been argued that the loss of constraint occurred in two phases, an early one, and another approximately 0.75 million years ago (Wang et al. 2004). The first is assumed to be incomplete, affecting the ancestry of humans and chimpanzees, while the second more complete relaxation affected the hominid lineage alone. Dietary changes occurred during this time span in hominid evolution. Two million years ago hominids increased their consumption of animal food and decreased their intake of plant food (Milton 2003) and therefore are likely to have reduced their exposure to a number of toxins. About 0.8 million years ago hominids invented the controlled use of fire (Goren-Inbar et al. 2004), the time corresponding well with the onset of the second wave of relaxation of selective constraint. Since cooking detoxifies poisonous food, hominids again reduced their exposure to toxins. Thus, these factors are likely to have reduced the selective pressure of human *TAS2R* genes (Wang et al. 2004). The observation that *TAS2R* genes are under relaxed selective constraint has several implications. The loss of genes in humans would decrease the number of bitter compounds that humans can taste. Second, the segregation of non-functional alleles in the current human population indicates inter-individual bitter taste variations. Third, new TAS2R alleles could appear that bind previously unrecognizable bitter substances, an event that, under neutral evolution, must not necessarily provide a fitness advantage (Wang et al. 2004).

Tissue distribution

Taste buds occur in the three types of lingual papillae and other parts of the oral cavity. The palate is particularly rich in taste buds in a region called "geschmacksstreifen." RT-PCR studies clearly demonstrated the presence of selected rodent and human TAS2R mRNAs in these taste tissues (Matsunami et al. 2000; Ueda et al. 2001; Bufe et al. 2002; Behrens et al. 2004). Fully consistent with the proposed role of the TAS2Rs as taste receptors subsequent in situ hybridization experiments unambiguously detected mRNAs for a large collection of rodent TAS2Rs in a subset of TRCs of all three types of lingual papillae and of the palate (Adler et al. 2000; Matsunami et al. 2000; Kim et al. 2003a). Similarly, the mRNAs for human TAS2R14, hTAS2R16, hTAS2R38, hTAS2R43, and hTAS2R44 were found in a subset of human circumvallate TRCs (Bufe et al. 2002, 2005; Behrens et al. 2004; Kuhn et al. 2004).

From the number of positive cells and experimental parameters, it has been estimated that 6–10 cells per taste bud express TAS2Rs, corresponding to approximately 15% of the cells. This estimate applies to rodent vallate, foliate, geschmacksstreifen, and epiglottis taste buds, where almost every taste bud contains TAS2R-positive cells (Adler et al. 2000). A similar estimate of approximately 20% has been reached for TAS2R gene expression in human circumvallate taste buds (Meyerhof et al. 2005). In marked contrast, less than 10% of rodent fungiform papillae contained TAS2R mRNA-positive cells, yet their number per positive taste bud did not differ from that of taste buds of the other oral regions (Adler et al. 2000). It is not clear at present whether this pattern of *TAS2R* gene expression generates a topographic map of different bitter taste sensitivities on the tongue (Lindemann 1999). Responses to various bitter compounds that have been recorded from the glossopharyngeal (GP) and chorda tympani (CT) nerves of mice (Danilova and Hellekant 2003) revealed a complex situation. The GP innervates the posterior tongue with the vallate and foliate papillae, whereas the CT innervates the anterior tongue. Although the regional distribution of fungiform papillae was difficult to assess, their density is apparently higher on the anterior tongue (Miller 1995). Recordings from the GP might therefore primarily reflect sensations from foliate and vallate papillae, while recordings from the CT reflect sensation from the fungiform papillae. Comparisons of the nerve responses recorded in C57BL/6J mice that were elicited by a number of bitter compounds revealed that, in general, the GP gave larger responses. However, responses to certain bitter compounds, including brucine, chloroquine, sparteine, quinine-HCL, and strychnine, were not significantly larger in the GP (Danilova and Hellekant 2003).

These observations have been confirmed in rats (Dahl et al. 1997). Similar recordings were also carried out in marmosets, which are New World monkeys (Danilova and Hellekant 2004). CT and GP fibers that responded best to bitter stimuli were quite different in their responses to the same set of bitter compounds. The responses to tannic acid and sucrose octaacetate were significantly larger in the CT fibers, whereas GP fibers showed larger responses to denatonium and caffeine. Also, the response characteristics were different. If these observations argue for a chemotopic map for bitter taste, it remains to be seen whether it can be transferred to humans; at present no such data exist for this species.

Further experiments demonstrated that mT2Rs did not colocalize with T1R sweet or umami taste receptors, but were exclusively expressed in α-gustducin-positive cells (Adler et al. 2000). In line with experiments demonstrating functional coupling of recombinant TAS2Rs to α-gustducin (Chandrashekar et al. 2000; Ueda et al. 2001) and with reports showing that bitter taste transduction involves this G protein subunit (Wong et al. 1996a, b; Ming et al. 1998), the obligatory expression of mT2Rs in α-gustducin-positive cells strongly underscores the role of oral TAS2Rs as bitter taste receptors.

Adler et al. (2000) also observed by in situ hybridization experiments that mixed probes of 2, 5 or 10 different mT2Rs did not label more TRCs than single probes, yet the signal intensities per cell were increased. These results did not differ between oral taste buds. From these data they concluded that each positive TRC expresses a large number of mT2Rs, possibly nearly the full complement. Colocalization experiments clearly confirmed the coexpression of pairs of mT2Rs in the same TRCs. Since mT2Rs signal through the same pathway (Zhang et al. 2003), these data suggest that there would be little functional discrimination between mT2R-positive cells. This conclusion was supported by behavioral experiments demonstrating that rats failed to discriminate denatonium from quinine (Spector and Kopka 2002) and that rhesus monkeys generalized quinine with strychnine, denatonium benzoate, phenylthiocarbamide (PTC), caffeine, and urea (Aspen et al. 1999). However, these findings have been challenged by calcium imaging experiments carried out on mouse isolated taste buds, which showed that the majority of TRCs in foliate slices responded to only one bitter compound out of five compounds tested (Caicedo and Roper 2001). At present, this contradiction cannot be resolved. Perhaps the calcium imaging data do not precisely reflect bitter taste receptor-expressing cells or cells that contribute to neurotransmitter release and thus generate a nerve signal. Perhaps mixed probes of 10 mT2Rs do not report the precise equipment of TRCs with bitter taste receptors; or perhaps interactions of mT2Rs present in the same subset of cells generate functionally distinct TRCs; perhaps the presence of mT2R mRNA in TRCs does not predict the presence of protein; or perhaps the concentrations of individual TAS2R mRNAs vary in different TRCs. Future experiments, in which the presence or absence of TAS2Rs with known bitter agonists can be assessed simultaneously with functional responses of individual TRCs to the respective bitter agonists in combination with rigorous behavioral discrimination tasks, might clarify this contradiction.

Recent studies identified TAS2Rs in tissues outside the oral cavity. In rats and mice, an extensive population of chemosensory cells has been found in the nasal epithelium that forms synaptic contacts with afferent trigeminal fibers. These cells express rodent T2Rs together with α-gustducin (Finger et al. 2003). Application of bitter compounds to the nasal epithelium activated the trigeminal nerve and induced changes in the respiratory rate, suggesting that the TAS2R-expressing cells are part of protective mechanisms, including apnea and sneezing, that remove inhaled irritating substances. Using RT-PCR, multiple rodent T2Rs have been detected in the antral and fundic gastric mucosa as well as in the lining of the duodenum and enteroendocrine STC-1 cells (Wu et al. 2002). In these tissues, the mRNA for α-gustducin was present as well. When STC-1 cells were challenged with denatonium, cycloheximide or PTC rapid increases in the cytosolic calcium concentrations were observed, suggesting that intestinal cells express functional TAS2Rs. The role of the gastric and intestinal mucosa in sensing the chemical composition of the gut lumen has been known for a long time, although the molecular mechanisms remained elusive. It has also been observed that denatonium stimulates release of insulin from pancreatic β-cells (Straub et al. 2003), suggesting that TAS2Rs are also expressed in this cell type. Collectively, the data may point to a more general role of TAS2Rs in chemosensation. In light of the apparently obligatory expression of TAS2Rs with α-gustducin and the fact that α-gustducin has been observed in several different tissues (Höfer et al. 1996; Adler et al. 2000; Kuhn et al., unpublished observations), TAS2Rs probably have many functions in addition to their role as bitter taste receptors.

Transduction mechanisms

The discovery of α-gustducin in the laboratory of Robert F. Margolskee was a milestone in the elucidation of taste transduction mechanisms (McLaughlin et al. 1992). This G-protein subunit, which is closely related to transducin, the G protein of visual sensory transduction, was specifically detected in a subset of taste receptor cells of rodents and humans (McLaughlin et al. 1992; Takami et al. 1994), making it an attractive candidate for transducing taste stimuli. In fact, the participation of α-gustducin in bitter taste transduction was suggested by the observation that various bitter stimuli activated this G protein α subunit in bovine taste membranes (Ming et al. 1998). These studies also showed that the carboxyl-terminal part of α-gustducin physically interacted with the unknown taste receptors, as peptides derived from regions of transducin known to interact with rhodopsin competitively inhibited activation of α-gustducin (Ming et al. 1998). Activation of α-gustducin through bitter taste receptors was quite recently verified by the demonstration that the recombinant mT2R5 stimulated the incorporation of radiolabeled GTP in insect cell membranes when exposed to its ligand cycloheximide (Chandrashekar et al. 2000). Moreover, in heterologous expression assays a number of hTAS2Rs coupled efficiently to a chimeric G protein containing the receptor-coupling sequences of α-gustducin at its carboxyl terminus (Ueda et al. 2003; Behrens et al. 2004; Kuhn et al. 2004; Bufe et al. 2005). Finally, gene-targeted mice devoid of a functional gustducin gene show impaired behavioral and electrophysiological responses to bitter stimuli (Wong et al. 1996a), while transgenic expression of rat α-gustducin restored responsiveness of gustducin null mice to bitter compounds (Wong et al. 1999). The role of α-gustducin was further established by a combination of immunohistochemistry and calcium-imaging experiments in mouse taste buds (Caicedo et al. 2003). These experiments revealed that many but not all mouse vallate TRCs that responded to bitter stimuli expressed α-gustducin. As expected, in similar experiments carried out using taste buds of gustducin null mice the incidence of cells responding to bitter stimuli was largely reduced. Yet some cells lacking α-gustducin responded to stimulation with bitter compounds. These experiments are remarkable because they offer an explanation for the phenotype of the gustducin null mice, which showed reduced but not abolished responses to bitter stimuli. The reduction in bitter taste sensitivity is likely to be due to a residual population of bitter responsive cells (Caicedo et al. 2003). Other G proteins probably couple to the taste receptors in these cells. Likely candidates are other Gi-type proteins, including transducin. Transducin is expressed in TRCs and can partially rescue gustducin null mice and functionally couple to bitter taste receptors (Ruiz-Avila et al. 1995; Yang et al. 1999; He et al. 2002). Moreover, in vitro studies showed that mT2R5 could also couple to pertussis toxin sensitive Gi/o type G proteins (Ozeck et al. 2004).

Although these studies clearly showed that α-gustducin plays a central role in transducing bitter taste and couples to bitter taste receptors, the coupling of α-gustducin to the intracellular effector molecules is less understood. Based on its pronounced sequence relationship to transducin, it has been proposed that α-gustducin, like transducin in photoreceptor cells, decreases the cyclic nucleotide concentration in TRCs (McLaughlin et al. 1994). This suspicion was supported by the identification of two types of phosphodiesterases in taste tissue (McLaughlin et al. 1994). Moreover, quench flow assays showed that stimulation of mouse taste tissue with the bitter compounds denatonium or strychnine rapidly and transiently reduced the levels of cyclic nucleotides. This sequence of reactions depended on α-gustducin, as the presence of a specific antiserum directed against this G protein blocked the effect (Yan et al. 2001). On the other hand this study and a number of other studies also showed that bitter stimuli resulted in a rise in inositol trisphosphate (IP$_3$; Akabas et al. 1988; Hwang et al. 1990; Spielman et al. 1994, 1996; Miwa et al. 1997; Nakashima and Ninomiya 1998) consistent with the observation that type III IP$_3$ receptor is coexpressed with α-gustducin and other signaling components of bitter taste transduction (Clapp et al. 2001; Miyoshi et al. 2001). The denatonium-mediated rise in calcium was blocked by U73122, a specific phospholipase C (PLC) inhibitor (Ogura et al. 1997). Subsequent studies found the PLC-β2 isoform to be specifically expressed in TRCs and showed that antibodies directed against PLC-β2 blocked the denatonium-mediated rise in IP$_3$ (Rössler et al. 1998; Yan et al. 2001). The central role of PLC-β2 was finally proven in gene-targeted mice. PLC-β2 null mice completely lost their ability to respond to bitter compounds (Zhang et al. 2003).

Since it was already known that PLC-β2 is regulated by G protein β/γ subunits (Katz et al. 1992), differential screening of cDNA libraries constructed from single α-gustducin expressing TRCs was carried out. This study identified G protein γ13, a novel G protein γ subunit distantly related to the previously known Gγ subunits (Huang et al. 1999). Gγ13 colocalized with α-gustducin in mouse TRCs. Further expression profiling led to the conclusion that the heterotrimeric gustducin complex consists of α-gustducin/Gβ3γ13 and/or α-gustducin/β1/γ13. Consistent with this assumption, antibodies directed against γ13, β1 or β3 blocked the denatonium-induced rise in IP3 in mouse taste tissue (Huang et al. 1999; Rossler et al. 2000).

To identify genes encoding further signaling components expressed in α-gustducin-positive TRCs, the same approach of differential screening single cell cDNA libraries identified a novel member of the transient receptor potential (TRP) channel, TRPM5 (Perez et al. 2002). Mammalian TRP channels form a large family of proteins that can be grouped into six subfamilies based on their sequence relationship (Clapham 2003). These channels play important roles in sensory transduction and respond to various stimuli, including heat and cold, noxious chemicals, hypo-osmolarity or GPCR activation. TRPM channels obtained their name from the first identified channel of this family, melastatin, a protein of unknown function that is upregulated in metastatic melanoma cells (Duncan et al. 1998). TRPM5 expression in mouse TRCs correlated with that of α-gustducin, Gβ3, Gγ13, and PLC-β2 (Perez et al. 2002). Moreover, TRPM5 knockout mice lost their electrophysiological and behavioral responses to various bitter stimuli making this channel a central player in bitter transduction (Zhang et al. 2003). Functional analyses of recombinant TRPM5 in oocytes (Perez et al. 2002) or mammalian cells produced conflicting results with regard to activation and conductance properties. Subsequent studies revealed that TRPM5 is a monovalent-specific cation channel directly activated by calcium ions at μM concentrations (Hofmann et al. 2003; Liu and Liman 2003; Prawitt et al. 2003). TRPM5 shows rapid deactivation upon receptor

                                      Rev Physiol Biochem Pharmacol (2005) 154:37–72

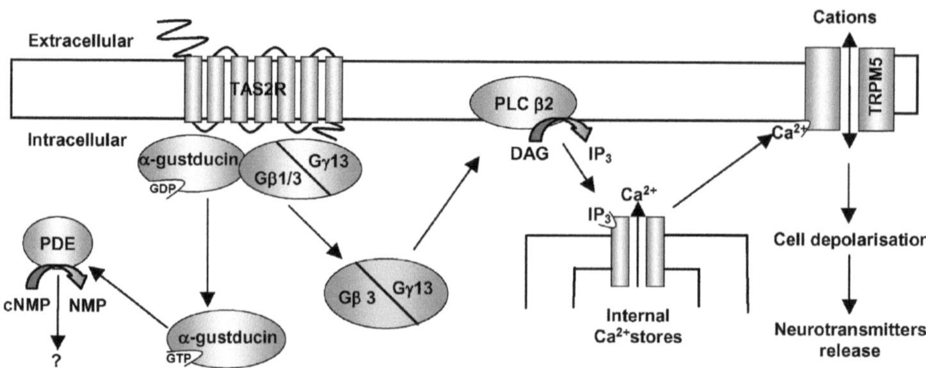

Fig. 3 Proposed bitter taste transduction in taste receptor cells (TRCs). *cNMP* cyclic nucleoside monophosphate, *NMP* nucleoside monophosphate, *PDE* phosphodiesterase, *PLC* phospholipase C, *IP3* inositol triphosphase

stimulation and is modulated by membrane potential, phosphatidylinositol-4,5-bisphosphate and intracellular calcium. Sustained exposure to calcium desensitized TRPM5 (Liu and Liman 2003). Moreover, TRPM5 required rapid changes in $[Ca^{2+}]_i$ to generate significant whole-cell currents, whereas slow elevations in $[Ca^{2+}]_i$ to equivalent levels were ineffective (Prawitt et al. 2003). This particular behavior may well account for the contradictory observations made in the previous studies (Perez et al. 2002; Zhang et al. 2003).

The above findings propose signal transduction pathways for bitter stimuli in TRCs (Fig. 3). Activated TAS2Rs couple to Gi-type proteins, the most likely candidate in most TRCs being gustducin, but transducin and/or Gi2 are alternatives or may do so in some TRCs. Dissociation of the heterotrimeric gustducin splits the signal into two parts. Alpha-gustducin decreases cyclic nucleotide levels. The β3/γ13 complex activates PLC-β2, a rise in IP3, and elevation of cytosolic calcium concentration. Calcium ions in turn stimulate TRPM5 channels resulting in changes of membrane currents leading to a receptor potential, which translates into action potentials and release of neurotransmitter. The consequences of the α-gustducin induced decrease in cyclic nucleotide levels are still not as well understood. Although the presence of phosphodiesterases and a cyclic nucleotide-gated (CNG) channel subunit have been described (McLaughlin et al. 1994; Misaka et al. 1997), their link to α-gustducin has not yet been established. Moreover, cyclic nucleotides have been reported to suppress membrane currents in frog taste cells (Kolesnikov and Margolskee 1995), whereas they would activate the CNG channel. Future work will elucidate the possible roles of the gustducin alpha subunit and cyclic nucleotides in bitter taste transduction in TRCs.

Structure–function relationships of TAS2Rs

A complete understanding of the physiology of bitter taste requires exact knowledge about the interactions of TAS2Rs with bitter substances. In other words, the tasks are first to deorphan all TAS2Rs, and second to elucidate their structure–activity relationships. These are great challenges for several reasons. Firstly, there is a vast excess of numerous, structurally diverse bitter compounds (Belitz and Wieser 1985; Keast and Breslin 2002) over the ~30 mammalian TAS2R bitter taste receptors. The TAS2Rs are therefore highly unlikely to interact with their bitter stimuli in a simple one-to-one ratio. They should be broadly tuned to multiple bitter compounds. If they are narrowly tuned to few or even to specific compounds, TAS2Rs would only mediate the bitter taste of a minority of bitter substances and additional detection mechanisms must exist for bitter taste. Secondly, current heterologous expression systems do not copy the signal transduction pathway of TRCs. Not all TAS2Rs may transduce binding of their agonists to signaling cascades in heterologous systems. Thirdly, chemosensory receptors are not commonly targeted to the plasma membrane in heterologous expression systems and might therefore be inaccessible for their agonists (Krautwurst et al. 1998; Wetzel et al. 1999; Chandrashekar et al. 2000).

Dealing with these potential problems three groups recently reported on the identification of bitter tastant–receptor combinations. Two of them used human embryonic kidney (HEK) 293 cells, a standard cell line widely used in molecular biology. They engineered them to express $G\alpha15$ to identify agonists for TAS2Rs (Bufe et al. 2002; Chandrashekar et al. 2000). This G protein promiscuously couples a variety of heptahelical receptors to the release of intracellular calcium (Offermanns and Simon 1995; Bockaert and Pin 1999). In such systems, receptor activation results in elevated levels of intracellular calcium, which can be monitored using calcium-sensitive fluorescence dyes. Based on recent observations (Krautwurst et al. 1998) showing that the amino-terminal amino acids of bovine rhodopsin targeted several odorant receptors to the plasma membrane of HEK293 cells, Chandrashekar et al. (2000) facilitated cell surface targeting of the recombinant TAS2Rs, by adding the first 39 amino acids of bovine rhodopsin to the amino-termini of various human and murine TAS2Rs. Subsequently, they challenged the transfected cells after loading with the calcium indicator Fura-2 with 55 different tastants and analyzed their responses by single cell calcium imaging, which provides highest sensitivity.

In similar investigations Bufe et al. attached the membrane targeting sequence of the rat somatostatin 3 receptor (Ammon et al. 2002) to the amino-termini of the recombinant TAS2Rs allowing their cell surface expression in HEK293 cells also expressing $G\alpha15$ (Bufe et al. 2002, 2004). These authors used Fluo-4 as the calcium indicator and an automated fluorometric plate reader that measures the calcium responses of the total population of cells present in the well of a microplate. Since untransfected cells contribute to the background fluorescence but do not respond to the stimulus, this method is somewhat less sensitive than single cell calcium imaging. However, due to the microplate format it allows a higher throughput. Later on, these authors increased the sensitivity of their assay (Behrens et al. 2004; Kuhn et al. 2004; Bufe et al. 2005) by replacing $G\alpha15$ with a chimeric G protein α subunit consisting of human $G\alpha16$ at the amino terminus, whereas the last 44 amino acids were from human α-gustducin (Ueda et al. 2003). The use of this G protein chimera also increased the physiological relevance of the assay, as G protein α subunits couple predominantly with their carboxyl-terminal parts to the receptors. The use of this chimeric G protein allowed more efficient coupling of the TAS2Rs to the calcium response.

An alternative strategy to deorphan hTAS2Rs (Pronin et al. 2004) relied on the observation that agonist stimulation of mT2R5 expressed in *Spodoptera frugiperda* SF9 insect membranes induced the incorporation of radiolabeled $GTP\gamma S$ into α-gustducin (Chandrashekar et al. 2000). However, because of its limited availability in pure form, these authors replaced α-gustducin with transducin, the visual G protein α subunit, in their biochemical investigation of hTAS2Rs. Transducin is present in high amounts in the retina, is structurally related to α-gustducin and has been implicated in taste signal transduction (Ruiz-Avila et al. 1995; Ming et al. 1998). The agonist-dependent activation of a taste tissue G protein is more likely to reflect the physiological situation than the $G\alpha15$-dependent system. Moreover, it circumvented the potential problem that not all TAS2Rs may couple to $G\alpha15$. On the other hand, this approach has the lowest throughput.

Table 3 Identified bitter compound/TAS2R combinations

Receptor	Identified agonists	Reference
mT2R5	Cycloheximide, lidocaine	(Chandrashekar et al. 2000; Pronin et al. 2004)
hTAS2R10	Strychnine	(Bufe et al. 2002)
hTAS2R14	α-Thujone, picrotoxinin, picrotin, 1-naphthoic acid, benzoate, piperonylic acid, 1-nitronaphthalene, 1,8-naphthalaldehydic acid	(Behrens et al. 2004)
hTAS2R16	Various β-glucopyranosides, including salicin, helicin, arbutin, phenyl-β-D-glucoside, methyl-β-D-glucoside, amygdalin, esculin, 2-nitro-phenyl-β-D-glucoside, naphthyl-β-D-glucoside	(Bufe et al. 2002)
hTAS2R38	Various thioamides, including PROP, PTC, diphenylthiurea, acetylthiourea, methylthiouracil	(Bufe et al. 2005)
hTAS2R43	Aristolochic acid, 6-nitrosaccharin, saccharin, acesulfame K, n-isopropyl-2-methyl-5-nitrobenzenesulfonamide	(Kuhn et al. 2004; Pronin et al. 2004)
hTAS2R44	Aristolochic acid, saccharin, acesulfame K,	(Kuhn et al. 2004)
hTAS2R47	Denatonium, 6-nitrosaccharin	(Pronin et al. 2004)

PROP propylthiouracil, *PTC* phenylthiocarbamide

Mouse T2R5 is the first TAS2R for which an agonist has been identified (Chandrashekar et al. 2000). Cells expressing this receptor responded dose-dependently to cycloheximide, a substituted ethylglutarimide antibiotic (Table 3). The responses occurred in nearly all transfected cells and depended critically on the presence of the transfected receptor and G protein. Mouse T2R5 displayed a half maximal response at approximately 0.5 μM (EC_{50}) and a threshold of activation of approximately 0.2 μM in the in vitro assay. The sensitivity of the receptor thus closely matched the sensitivity of tasting cyloheximide in mice, which are strongly aversive to this compound and display a sensitivity threshold of approximately 0.25 μM (Lush and Holland 1988). The response of mT2R5 initially appeared to be quite selective, as none of the other 55 different taste compounds elicited a cellular response, even when administered at much higher than the biologically relevant concentrations. This result was reproduced with the rat homolog, rT2R9 (Bufe et al. 2002). However, later on it was discovered that mT2R5 also responded to lidocaine (Pronin et al. 2004), a synthetic derivative of the alkaloid cocaine, which is used as a class IB antiarrhythmic and local anaesthetic. In similar assays hTAS2R4 and its mouse homolog mT2R8 recognized the synthetic compound denatonium benzoate and the unrelated thioamide PROP (Chandrashekar et al. 2000). They did so, however, only at very high concentrations, several orders of magnitude above the human detection thresholds, suggesting the existence of additional agonists with higher potency for hTAS2R4 and mT2R8. Taken together, these receptors recognize chemically distinct compounds within different concentration ranges.

Human TAS2R16 recognized several β-glucopyranosides, all of which were bitter (Bufe et al. 2002). However, their threshold values of activation and EC_{50} values differed approximately 150-fold, but were similar and showed the same rank order in vivo and in vitro. These compounds have an aglycon attached via a β-glycosidic bond to the pyranose ring. Structure–activity studies revealed structural parameters that made compounds agonists for hTAS2R16 (Fig. 4). The β-configuration of the glycosidic bond was indispensable, as α-glucopyranosides did not activate the receptor. The steric position of the hydroxyl group at the C4 carbon atom of the pyranose was also crucial; only β-glucopyranosides activated the receptor, but not a β-galactoside. The aglycons present at C1 exerted important but not es-

	compound	structure	EC_{50} (mM)
hTAS2R16	phenyl-β-D-glucopyranoside	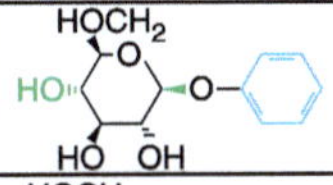	1.1 ± 0.1
	arbutin	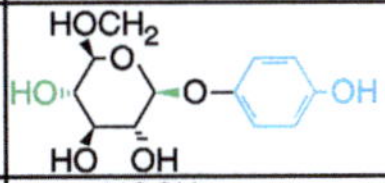	5.8 ± 0.9
	methyl-β-D-glucopyranoside	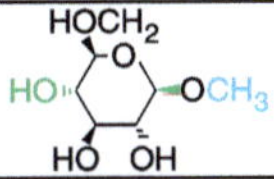	>5 0
	phenyl-β-D-galactopyranoside	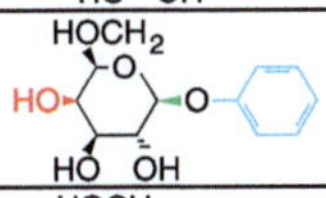	no response
	phenyl-α-D-glucopyranoside	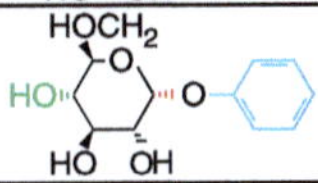	no response
hTAS2R38	phenylthiourea	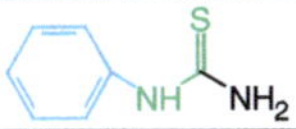	1.1 ± 0.5
	acetylthiourea	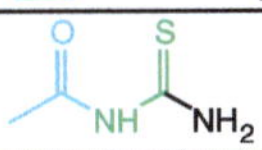	25 ± 16
	phenylurea	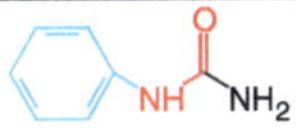	no response
	propylthiouracil	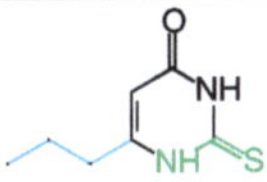	2.1 ± 0.9
	methylthiouracil	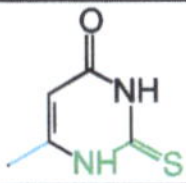	58 ± 38
	thiobarbituric acid	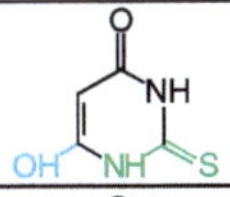	no response
	uracil	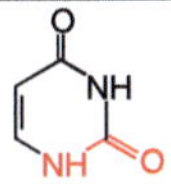	no response

Fig. 4 Structure–activity relations of some compounds to hTAS2R16 or hTAS2R38. EC_{50} values were derived from concentration–response curves recorded in cells expressing either of the two receptors following bath application of the compounds at various concentrations (Bufe et al. 2002, 2005). Substructures shown in *green* are crucial for receptor activation; substructures in *blue* modify the efficacy of activation. Substructures shown in *red* do not allow activation of the receptors

sential effects on receptor activation. Aromatic phenyl rings had better agonistic properties than a small methyl group. Hydrophilic substitutions in the ring systems reduced the agonist strength (Fig. 4). Substitutions at C6 had only a small effect on receptor activation; the moderate potency of amygdalin was better explained by the presence of a hydrophilic nitrile group in the aglycon than by the α-1,6-glycosidic addition of another pyranose ring. Data characterizing the bitter taste of glucopyranosides supported these findings (Kubo 1994).

From this study, hTAS2R16 appears to be tuned to various β-glucopyranosides, and the data predict that many more than the tested substances will activate the receptors provided they fulfill the above requirements. Considering that some 10% of plants contain toxic glycosides (Kingsbury 1964), the number of potential hTAS2R16 agonists may be enormous.

hTAS2R38 showed similar tuning characteristics (Bufe et al. 2005). This receptor was tuned to chemicals containing thioamide moieties, the most known representatives being PTC and PROP (Fig. 4). Chemicals with amide groups did not activate hTAS2R38. The thioamide moiety occurred in various chemically different agonists. However, the potency of the hTAS2R38 agonists was also influenced by the presence of other chemical groups near the thioamide substructure. The presence of less hydrophobic or even hydrophilic groups dramatically diminished agonist properties. Replacing the propyl moiety of PROP with a hydroxyl group gives thiobarbituric acid, a compound that failed to activate hTAS2R38. Like hTAS2R16 agonists, hydrophilic substitutions in hTAS2R38 agonists clearly diminished their activation abilities. So far, only a limited number of thioamides have been shown to activate hTAS2R38. However, since this receptor determines the bimodal taste sensitivities of humans to PTC and many more compounds (Harris and Kalmus 1949; Barnicot et al. 1951), the number of agonists for hTAS2R38 is certainly much larger.

Human TAS2R14 appeared to be surprising (Behrens et al. 2004), in that it responded to almost a quarter of the test substances, i.e. eight out of 33 different bitter compounds (Table 3). The most powerful agonists for hTAS2R14 were the structurally diverse compounds α-thujone, picrotoxinin, and naphthoic acid. There appeared to be no obvious common structural motif shared by all hTAS2R14 agonists except that they contain one or several ring systems and at least one electronegative side chain, which were, of course, also found in many bitter compounds that did not activate hTAS2R14. Moreover, hTAS2R14 agonists varied considerably in size, the three most powerful agonists containing 1, 2 or 4 ring systems. These results do not mean that hTAS2R14 displayed little specificity for its agonists as relatively small differences between the molecular structures of picrotoxinin and picrotin or of benzoate, naphtholic acid, and nitronaphthalene had a clear impact on receptor activation.

Human TAS2R43 and hTAS2R44 belong to group A TAS2Rs (Shi et al. 2003) and show a pronounced sequence relationship. They are the two closest related paralogs of the entire human TAS2R family showing 89% amino acid identity (Bufe et al. 2002). Both receptors have been shown to mediate the bitter aftertaste of sulfonyl amide sweeteners (Kuhn et al. 2004). Saccharin and acesulfame K have an intrinsic bitter aftertaste to some people, which increases with concentration and limits the use of these compounds (Schiffman and Gatlin 1993; Horne et al. 2002). Human TAS2R43 and hTAS2R44 responded with calcium signals in transfected cells to saccharin and acesulfame K as well as to the purely bitter tasting natural compound, aristolochic acid (Kuhn et al. 2004). Aristolochic acid is the primary bitter principle of *Aristolochia* species and present in many oriental remedies (Ganshirt 1953). However, aristolochic acid was much more potent at both receptors. Its concentration–response curves were shifted leftward to lower agonist concentrations by approximately 6 and 3 orders of magnitude relative to those of the sulfonyl amides at hTAS2R43 and hTAS2R44 respectively. Moreover, the two receptors differed in their activation behavior; the EC_{50} values for aristolochic acid was approximately 5-fold lower at hTAS2R43 and this compound also had a lower efficacy at hTAS2R44. *Vice versa*, the two sulfonyl amides were more potent at hTAS2R44. It was also found using the GTPγS binding assay that hTAS2R43 responded to the saccharin analog 6-nitrosaccharin and the related compound N-isopropyl-2-methyl-5-nitrobenzenesulfonamide (IMNB), a compound that differs from 6-nitrosaccharin by the open ring structure (Table 3; Pronin et al. 2004). In this study, saccharin itself did not activate hTAS2R43, most likely reflecting the lower sensitivity of the biochemical assay. 4-Nitrosaccharin and p-nitrotoluene, which are not bitter to humans, were unable to stimulate GTPγS binding (Hamor 1961; Pronin et al. 2004). From these observations, it appears that the nitrobenzene group is not sufficient to activate the two receptors, but its presence in an appropriate orientation enhances agonist potency. The crucial parameter for activation of hTAS2R43 and hTAS2R44 appeared to be the negatively charged sulfonylamide or carboxyl groups.

Human TAS2R47 is also a member of group A TAS2Rs and closely related to hTAS2R44. Like hTAS2R43, hTAS2R47 also recognized 6-nitrosaccharin, although less efficiently (Pronin et al. 2004). Human TAS2R47, however, did not respond to IMNB suggesting that they share determinants mediating the activation of hTAS2R43, but distinct properties of IMNB are required for stimulation of hTAS2R47. The data obtained with hTAS2R43, hTAS2R44, and hTAS2R47 also support the idea that closely related TAS2Rs have an overlapping set of bitter agonists. Human TAS2R47 was also activated by μM concentrations of denatonium benzoate. Pronin et al. (2004) demonstrated that the denatonium moiety was responsible for this effect. Denatonium analogs that elicit reduced bitterness in humans also less potently activated hTAS2R47 with the same rank order of potency in vivo and in vitro (Saroli 1984; Pronin et al. 2004). Several conclusions can be drawn from these data. First, at least two receptors contribute to the bitter taste of denatonium benzoate. Human TAS2R47 is likely to mediate the bitter taste of the denatonium moiety (Pronin et al. 2004), while hTAS2R14 is responsible for the bitter taste of the food preservative benzoate (Behrens et al. 2004). Human TAS2R4, which is activated at denatonium concentrations about 5 orders of magnitude above the human threshold for this compound, may be a low affinity receptor for either denatonium or benzoate.

Only one study examining the interactions of bitter compounds with TAS2Rs by mutational analysis has been published so far (Pronin et al. 2004). The authors generated chimeric receptors between the closest related TAS2Rs, hTAS2R43, which responded in the biochemical assay to 6-nitrosaccharin and IMNB and hTAS2R44, which did not. Sequence comparisons revealed that 15 out of 34 different amino acid positions are present in the extracellular loops (EL)-1 and EL-2, while EL-3 is completely conserved. Swapping EL-1, which differs in only four positions between the two receptors, rendered the variant hTAS2R43 insensitive to IMNB, while the variant hTAS2R44 acquired sensitivity to IMNB. Swapping EL-2 had no effect on IMNB activation. While other parts of the receptor are also likely to interact with IMNB, EL-1 appears to be crucial for activation by IMNB. Interaction of hTAS2R43 with 6-nitrosaccharin was more complex. Swapping both EL-1 and EL-2 made the hTAS2R44 variant sensitive, whereas the hTAS2R43 variant lost most but not all of its sensitivity. The latter observation indicates that also other parts of hTAS2R43 are interacting with 6-nitrosaccharin. Swapping either loop made all constructs responsive to 6-nitrosaccharin, while the larger effect was contributed by EL-1.

Analysis of single point mutations further revealed that exchanging G92 in EL-1 of hTAS2R44 for the corresponding asparagine residue of hTAS2R43 made the mutant responsive to both ligands. The reciprocal substitution in hTAS2R43 largely impaired its activation properties, suggesting that N92 plays a critical role in the activation of hTAS2R43. The importance of N92 was also underscored by the observation that an aspartate in this position increased the receptor's basal activity about 4-fold. When W88, a residue present in EL-1 in 21 out of the 25 hTAS2Rs, was mutated to the small aliphatic amino acid alanine hTAS2R43 completely lost its ability to become activated. Exchange of this residue by phenylalanine had virtually no effect. Human TAS2R47 was similarly sensitive to these mutations. N92G and N92A rendered this receptor unresponsive, while the N92D variant retained marginal sensitivity. The mutation W88F also completely abolished activation of hTAS2R47, whereas it had no effect on hTAS2R43. These results do at present not allow precise roles to be assigned to the amino acids of EL-1. They might be involved in ligand binding interactions, forming the binding pocket, or globally changing the receptor conformation.

Collectively, the results obtained from studying ten TAS2Rs so far allow limited insights into the interactions of TAS2Rs and their bitter agonists. In accordance with the expectation that TAS2Rs recognize multiple compounds, one important observation from these studies was that, in fact, TAS2Rs are quite broadly tuned to different bitter chemicals. In all cases but one, the TAS2Rs responded to various bitter compounds. In the exceptional case of hTAS2R10 for which only strychnine arose as a single agonist (Bufe et al. 2002), the author predicts that in the near future, as deorphaning TAS2Rs proceeds, additional agonists will be identified.

Apparently, hTAS2R16 and hTAS2R38 turned out to be receptors recognizing distinct substructures, i.e. bitter β-glucosides and thioamides. Whether other receptors share this behavior remains to be seen. In the case of the other TAS2Rs, deorphaned to date, it has not been possible to identify common structural properties among the multiple structurally unrelated compounds that activated the same TAS2R. This behavior appears to be unique for or at least a peculiarity of TAS2Rs amongst class A GPCRs. TAS2Rs may not have a narrow binding pocket but instead a binding region able to accommodate structurally divergent compounds. Alternatively, they may have several or overlapping binding pockets, each responsible for the interactions with a specific compound or a group of related compounds. The analysis of hTAS2R43 supports this idea, as the mutations had different effects on IMNB and 6-nitrosaccharin sensitivities (Pronin et al. 2004).

It appears somewhat contradictory in view of the ability of TAS2Rs to recognize structurally different compounds that they also show clear specificity. This is evident from the fact that all deorphaned receptors have also been tested with a collection of other bitter compounds to which they did not respond. Moreover, relatively small structural variation had a severe impact on receptor activation. For example, only β-glucosides, but not an α-glucoside or a β-galactoside activated hTAS2R16. Similarly, human TAS2R43 recognized 6-nitrosaccharin, but not 4-nitrosaccharin, 6-amino-saccharin or 6-carboxy-saccharin. Thus, precise interactions of the chemical groups of agonists with the peptide backbone and/or amino acid side chains are required to activate the TAS2Rs.

Another interesting aspect is that the closely related hTAS2R43 and hTAS2R44, as well as hTAS2R43 and hTAS2R47, shared some of their agonists, although their complete agonist spectrum was not identical. This observation raises an interesting point. Do the agonist spectra overlap only among very closely related TAS2Rs or is this a common property of many TAS2Rs? The answer is pending and can be given only when more TAS2Rs have been deorphaned or when those that are available today have systematically been cross-checked for the complete set of available bitter compounds.

Although quite a number of TAS2Rs have been deorphaned, the present heterologous expression systems differ from the TRCs, i.e. the physiological site of expression. First, TRCs have contact to bitter tastants only at their apical part, while the basolateral part is shielded from them. In contrast, bitter tastants have access to the entire cell surface of transfected mammalian cells. Many bitter tastants combine lipophilic and hydrophobic properties and thus have detergent-like properties. These properties, and the fact that various bitter tasting compounds are pharmacologically active substances, lead to cellular effects that influence the calcium levels in the cells. In this way, such bitter compounds generate artificial calcium signals in the absence of recombinant receptors (our own unpublished observations). Another problem may arise from the fact that the heterologous expression systems rely on the presence of coexpressed Gα15 or G protein chimeras. This aberrant G protein coupling could alter the agonist spectrum of a given receptor as has been observed for olfactory receptors (Shirokova et al. 2005). Another problem may come from the fact that tastants physiologically dissolve in saliva, while buffer/salt solutions are used to administer tastants to recombinant receptors. The perireceptor milieu could also have impact on receptor activation by bitter tastants (Kock et al. 1994; Schmale and Bamberger 1997). A shortcoming is also the different expression levels and extents of cell surface expression of the various TAS2Rs (Bufe et al. 2004), although we did not observe a direct correlation of expression levels and signal amplitude (Bufe et al., unpublished observations). Related to that point is the presence of membrane targeting sequences from other membrane proteins. It cannot be excluded today that these alter the activation properties of TAS2Rs. The solution of the problem will be to develop expression systems that are closer to the physiological situation. Reconstitution of the signal transduction pathways active in TRC, based on α-gustducin/β1 or β3/γ13, and PLC-β2 in the heterologous cell types, is one option. Driving cell surface expression of untagged TAS2Rs in transfected cells through coexpression of chaperone-like proteins that are unambiguously present in TRC and await identification is another option to improve the quality and reliability of functional expression systems. Such proteins were most recently discovered aiding the in vitro expression of recombinant olfactory receptors (Saito et al. 2004). These proteins, which belong to two protein families, called RTP and REEP, did not, however, improve the functional expression of the mouse T2R5. Whether this effect was specific for a particular bitter taste receptor or applies to the whole TAS2R family remains to be seen.

Experiments with transgenic animals prove the function of TAS2Rs bitter taste receptors

Mice and humans differ in their sensitivities to various bitter substances, including various β-glucopyranosides and PTC. These compounds elicit strong bitter taste in humans, whereas mice are largely indifferent to them. Transgenic expression exclusively in bitter TRCs of the human *TAS2R16* or human *TAS2R38* genes under a mouse T2R promoter enabled mice to detect and avoid phenyl-β-glucopyranoside and PTC at concentrations corresponding to the human taste sensitivity range (Mueller et al. 2005). These results validate the role of TAS2R16 and hTAS2R38 as bitter taste receptors for β-glucopyranosides and PTC in vivo. They also show that these receptors are sufficient to mediate the bitter taste of these compounds. Moreover, they indicate that differences in bitter taste across species are based on sequence differences in the respective TAS2R repertoire. To further examine the importance of TAS2Rs researchers knocked out the gene for the mT2R5 cycloheximide receptor. *T2R5* null mice dramatically lost their sensitivity to this highly aversive substance, even at concentrations approximately 100-fold higher than those required to trigger avoidance. The animals, however, retained their full responsiveness to other taste qualities and also to all the other bitter compounds tested (Mueller et al. 2005). Together these results prove that defined TAS2Rs are both necessary and sufficient for bitter taste perception.

Taste quality coding

A long-standing controversy in taste research relates to the question of how taste qualities are encoded. Two opposite theories exist, that may not be mutually exclusive but overlap

(Smith and St John 1999; Scott and Giza 2000; Katz et al. 2002). In its extreme, the labeled line (LL) theory assumes that the transmission of gustatory information proceeds via separate channels. Thus, specialized TRC populations exist for each of the basic taste qualities. These are innervated by dedicated fibers connecting them to separate populations of neurons. Their activities represent the primary taste qualities in the various brain areas. In this model, perception of bitter taste would result from the activation of a subset of TRCs in the oral cavity that are specifically tuned to detect bitter substances. Similarly, sweet compounds would trigger the activation of another subset of TRCs mediating the perception of sweetness. In the competing across-fiber pattern (or across-neuron pattern) theory (AFP) peripheral TRCs are broadly tuned to tastants of multiple taste qualities. They are connected to a population of neurons, such that they all contribute equally to the quality code. In this scenario each neuron, whether firing or not, contributes to an activity pattern generated by a given taste stimulus. Eventually, perception of a taste quality is achieved through the brain by decoding and interpreting these activity patterns. In this model, a bitter compound would activate a population of TRCs, the individual members responding with different strength. The population of TRCs activated, let us say, by a sour stimulus, overlaps with the previous one and may also differ in response amplitudes.

Both models are supported by experimental data. In favor of the AFP theory are electrophysiological recordings in rodents that consistently showed that TRCs, afferent nerve fibers, and neurons respond to stimuli of several taste qualities, although the response to one quality may be particularly strong (Dahl et al. 1997; Nishijo and Norgren 1997; Sato and Beidler 1997; Lundy and Contreras 1999; Gilbertson et al. 2001; Verhagen et al. 2003). Moreover, calcium imaging experiments in isolated mouse taste buds supported the broad tuning of TRCs (Caicedo et al. 2002). A broad tuning of neurons in the nucleus tractus solitarius was also seen in adaption and cross-adaptation experiments. In these experiments, adaptation to quinine cross-adapted to sucrose and adaptation to HCl cross-adapted to quinine in more than 50% of the cases (Di Lorenzo and Lemon 2000). On the other hand, recordings in primates showed a relatively narrow tuning of nerve fibers to bitter, sweet, and salty, supporting the LL theory (Hellekant et al. 1998; Danilova and Hellekant 2004). Support may also come from experiments that analyzed the induction of the immediate early marker cFOS in neurons of the rat nucleus tractus solitarius in response to taste stimulation. These studies revealed that a bitter and a sour stimulus elicited distinct patterns of cFOS immunoreactivity and that diverse bitter stimuli elicited highly similar patterns of cFOS immunoreactivity (Travers 2002; Chan et al. 2004) consistent with the existence of separate neuronal populations activated by bitter or sour stimuli and a common population of neurons responding to various stimuli of the same taste quality.

Cloning of taste receptors for sweet, bitter, and umami as well as of other taste signaling proteins now enabled researchers to address the problem of quality coding in the periphery. Three genes, *TAS1R1*, *TAS1R2*, and *TAS1R3* encode sweet and umami receptors. By heteromerization of the encoded proteins TAS1R1 and TAS1R3 form a functional receptor for umami, the taste of monosodium glutamate (Li et al. 2002; Nelson et al. 2002), whereas TAS1R2 and TAS1R3 assemble into a functional sweet taste receptor activated by all sweeteners tested so far (Nelson et al. 2001; Li et al. 2002; Jiang et al. 2005). TAS1Rs are expressed in a subset of about 30% of the TRCs. Their gene expression defines two major population of cells, one that coexpresses TAS1R1 and TAS1R3 and another that coexpresses TAS1R2 and TAS1R3 (Max et al. 2001; Montmayeur et al. 2001; Nelson et al. 2001). This pattern strongly suggests the existence of a distinct TRC population detecting sweet and umami stimuli. In addition, a minor population of TRCs has been observed expressing only TAS1R3 (Nelson et al. 2001). Largely, the expression of TAS1Rs does not overlap with

TAS2R expression, which is present in about 15% of vallate taste buds (Adler et al. 2000), consistent with the assumption that bitter, sweet, and umami TRCs are different. However, it should be noted here that another group observed a number of cells coexpressing TAS1R1 and TAS2Rs (Kim et al. 2003a). This apparent contradiction has not yet been clarified and its solution awaits future investigation. Since TAS2R expression is obligatorily coupled to the expression of α-gustducin and α-gustducin-positive cells are distinct from TRCs expressing hyperpolarization and cyclic nucleotide-gated channels 1 and 4, which is a candidate sour taste "receptor" (Stevens et al. 2001), it appears that there is a fourth separate cell population responding to sour stimuli. The signaling molecules PLC-β2 and TRPM5 are virtually coexpressed in 50% of the TRCs and the subsets of these cells express TAS1Rs or TAS2Rs, suggesting that TRCs tuned to different taste modalities share common signal transduction molecules (Miyoshi et al. 2001; Zhang et al. 2003). Together, these results are more likely to be compatible with the LL hypothesis.

Gene-targeting experiments in mice taught some very important lessons. As expected, all TAS1R null mice maintained their ability to respond to sour, salty and bitter stimuli (Damak et al. 2003; Zhao et al. 2003). TAS1R1 and TAS1R3 knockout animals were insensitive in behavioral and electrophysiological experiments to the umami stimuli. Similarly, TAS1R2 and TAS1R3 null mice lost most of their responses to sweet stimuli. While D-amino acid and artificial sweeteners were ineffective, these animals retained residual responses to high concentration of sugars (Damak et al. 2003; Zhao et al. 2003). The latter observation was consistent with the observation that in some TRCs TAS1R3 was detected in the absence of either TAS1R1 or TAS1R2 (Nelson et al. 2001). These residual responses have been explained through the activity of TAS1R homomeric receptors. TAS1R3 homomers respond to high sugar concentration in functional expression systems and a TAS1R2/TAS1R3 double knockout was completely blind to all sweet stimuli, including high concentrations of sugars (Zhao et al. 2003). Thus, sweet and umami taste strictly depend on different combinations of TAS1R receptors.

Mice with deleted TRPM5 or PLC-β2 genes lost their responses to sweet, umami, and bitter stimuli, while they responded normally to sour and salty stimuli demonstrating that sweet, bitter, and umami transduction employ different receptors, but converge on common crucial signaling molecules (Zhang et al. 2003). These animals provided a unique opportunity to functionally investigate taste quality coding in the periphery. Transgenic expression of PLC-β2 driven by the *mT2R5* gene promoter specifically rescued bitter taste responses, whereas these mice did not respond to sweet and umami stimuli. PLC-β2 function has also been rescued using two other promoters, namely those of the *mT2R19* and *mT2R32* genes. Remarkably, all three transgenic lines acquired full responsiveness to a large set of structurally diverse bitter compounds (Mueller et al. 2005). This result elegantly confirms that in mice most TAS2Rs are expressed in the same set of bitter TRCs. If these cells expressed only a subset of TAS2Rs, the transgenic lines would respond to discrete repertoires of bitter substances. Thus, bitter responsive cells appear to be broadly tuned to detect numerous bitter compounds. On the other hand, these data also clearly show that bitter receptor-expressing cells mediate responses to bitter but not to sweet and umami stimuli. Moreover, it appears that aversive responses to bitter compounds do not require the functionally attractive pathways of umami and bitter taste. Therefore, bitter is encoded independently of sweet and umami and TRCs are not tuned across umami, sweet and bitter taste qualities.

The links of sweet tastants to attractive pathways and behavior and the links of bitter tastants to avoidance pathways and behavior have been studied in greater detail. If the behavioral responses of taste stimulation were properties of the TRCs and not properties of the receptors they express, this would support the LL theory. The laboratories of Nicholas Ryba and Charles Zuker examined this problem in a series of pioneering experiments by reintroducing receptors in defined TRC populations. First, they employed a modified κ-opioid receptor, RASSL, (receptor activated solely by synthetic ligands; Redfern et al. 1999) for transgenic expression in TAS1R2-expressing cells using an inducible promoter. Wild-type animals or uninduced transgenic animals were completely insensitive to RASSL agonists. However, after induction of the transgene mice acquired a strong attraction to nanomolar concentrations of spiradoline, a RASSL agonist (Zhao et al. 2003). From their observation, the authors concluded that activation of the T1R2-expressing cells, rather than the receptors they express, i.e. the sweet taste receptor, determines behavioral attraction in mice. Thus, activation of a single cell type is sufficient to trigger specific taste responses. A combinatorial or temporal activation pattern is not needed to generate attraction in response to excitation of TAS1R2-expressing cells. Next, the inducible RASSL transgene was introduced into bitter taste cells under a T2R gene promoter (Mueller et al. 2005). Wild-type and uninduced transgenic mice were completely insensitive to spiradoline, even at high concentrations. Following induction of the transgene, however, mice showed a strong aversion to the RASSL agonist. This result impressively demonstrates that the same compound can induce strong attraction or aversion dependent on the cell type in which its receptor is expressed. This conclusion was further convincingly supported by the inducible expression of the *hTAS2R16* bitter receptor gene in TAS1R2-expressing cells. Induced animals only were now strongly attracted to phenyl-β-glucopyranoside, a compound they had avoided when the transgene was expressed in TAS2R-expressing cells (Mueller et al. 2005). Together, these investigations substantiate that sweet and bitter pathways are encoded peripherally by dedicated or labeled lines. The equipment of TRC with receptors thus determines the spectra of chemicals to which they respond. In this sense the receptors translate chemical structures into perception. The use of inducible promoters that rescued taste signaling pathways only in adult animals long after the taste system completed its developmental and wiring program also shows that the receptor cells for bitter and sweet appear to be hard wired and can be established without sensory input (Mueller et al. 2005).

How can we reconcile the results from these transgenic and knockout experiments pointing to defined TRC populations devoted to the three basic taste qualities, sweet, umami, and bitter, with the electrophysiological recordings and imaging experiments of TRCs suggesting the existence of TRCs that are broadly tuned across these taste qualities? One possibility to explain the contradiction is to assume that recording and calcium imaging experiments are more or less invasive, partially disrupting the tissue and exposing TRCs to taste stimuli also at their basolateral part, although this is not immediately evident in those experiments employing tissue slices (Caicedo et al. 2002). This could make the TRCs respond unphysiologically. Another possibility is to argue that the quite high concentrations of tastants stimulate more than one taste quality. This is well established in the case of the artificial sweetener saccharin, a compound that induces both sweet (Nelson et al. 2001; Li et al. 2002) and bitter taste receptors in similar concentration ranges (Kuhn et al. 2004). Moreover, some frequently used taste compounds exert pharmacological effects and may, based on this property, elicit artificial responses from taste receptor cells. Caffeine and theophylline modulate phosphodiesterases, nicotine is a long-known acetylcholine receptor agonist, and quinine acts on nucleic acid metabolism. But there are many more examples. Yet another possibility is to assume that the cells, which showed up in the electrophysiological or in the calcium imaging experiments, do not make synaptic contact with the afferent nerves. Perhaps they are immature cells, not yet fully differentiated. However, there is another possibility to explain the contradiction. Recently, researchers observed that taste bud cells expressing PLC-β2 and IP3 receptors correspond to the so-called type II cells, which can be identified by morphological parameters (Clapp et al. 2004). Since, sweet, umami, and bitter transduction relies on these signaling molecules they concluded that sweet, umami, and bitter TRCs may be type II cells. This conclusion is supported by the observation that type II cells also express α-gustducin in rat and bovine (Yang et al. 1999; Tabata et al. 2003). However, type II cells form no conventional synapses, which have been detected only in type III cells, another type of morphologically distinct taste bud cells (Miller 1995; Clapp et al. 2004). Type III cells also express voltage-gated calcium channels consistent with the presence of synapses, but the α-gustducin-expressing type II cells do not (Medler et al. 2003). In light of these data, it would be possible that TRC receptors expressing sweet, umami, and bitter receptors and synapse-containing cells in the taste bud form distinct populations. In this scenario, TRCs could relay signaling to type III cells, which excite the afferent nerves. In this case, receptor-expressing cells, as targeted in the gene knockout experiments, would converge on and excite mediator cells as observed in the recording and calcium imaging experiments. Alternatively, type II cells could excite the afferent nerves in an alternative fashion. Nonetheless, future work is required to elucidated intra-taste bud communication, identifying the receptor repertoire of the TRCs under study in electrophysiological or calcium imaging experiments before this issue will be clarified.

Perceptual differences and TAS2R gene polymorphisms

Bitter taste perception is highly individualized in humans (Bartoshuk 2000) and differs among strains of rats (Tobach et al. 1974). These taste differences may influence the diet and health (Tepper 1998) and have therefore attracted much attention. In humans the best-studied example of taste variation is the tasting of PROP and PTC (Guo and Reed 2001). The abilities to taste these two bitter compounds are bimodally distributed in humans, and thresholds of tasting PROP and PTC differ about 1000-fold between sensitive and insensitive subjects. The abilities to taste PROP and PTC are correlated and probably based on the same hereditary trait. Recently, the perceptual differences of humans in tasting chloramphenicol has also been shown to be of genetic origin; insensitivity to this compound is independent from PROP/PTC taste blindness and inherited by an autosomal recessive trait (Sugino et al. 2002). Similarly, taste differences in rodents to sucrose octaacetate, cycloheximide, raffinose undecaacetate, and quinine is genetically determined and the loci that determine tasting these compounds have been mapped on mouse chromosome 6 (Lush 1981, 1984, 1986; Lush and Holland 1988). Following identification of mT2R5 as the receptor for cyloheximide (cyx; Chandrashekar et al. 2000; Mueller et al. 2005) the hypothesis has been tested whether the gene for this receptor corresponds to the cyx locus. In fact, tight linkage between the *mT2R5* gene and the cyx locus was observed in crosses of cyx taster DBA/2J and cyx nontaster C57BL/6J mice. Sequence comparisons of the *mT2R5* genes isolated from various taster and nontaster strains revealed that all taster and nontaster strains shared the same mT2R5 alleles, respectively. The non-taster allele carried missense mutations changing the polypeptide sequence in five positions in the lower part of TM2, the upper part of TM3, EL-1, EL-2, and the carboxyl-terminal tail. Since these regions may be involved in cyx binding or G protein coupling, the sequence variations predicted functional differences. These differences have finally been confirmed by heterologous expression. The cyx dose–response curve for mT2R5 in nontaster mice was shifted rightward compared with that in taster mice reflecting the taste differences for cyx in these animals (Chandrashekar et al. 2000). These findings give first clear evidence that sequence variations in a taste receptor gene generates receptor variants with altered sensitivities that can be measured in vitro. These sensitivity differences determine perceptual differences in tasting a specific compound and eventually behavioral differences in avoiding this compound. Moreover, the above findings also predict that the genetically determined differences among mice strains in tasting sucrose octaacetate, raffinose undecaacetate, and quinine is due to sequence variations in other members of the mT2R family. They also suggest that perceptual variations in the human populations could have the same causes.

Similar findings in humans supported this idea. Genetic analysis of a complex trait in more than 250 members of 26 large three-generation Utah families identified a major locus for PTC tasting in an approximately 4 Mb region on chromosome 7q (Drayna et al. 2003). Analysis of the *TAS2R* genes by sequencing within families showing linkage to chromosome 7q identified numerous sequence variants, one of which demonstrated a strong association with the taste phenotype across the families analyzed. Further fine mapping of the chromosomal segment in question identified a small region of approximately 30,000 bp containing the *hTAS2R38* gene as the single gene (Kim et al. 2003b). It was further demonstrated that three single nucleotide polymorphisms (SNPs) in this gene defined five haplotypes encoding receptor variants differing in amino acid positions 49, 262, and 296. Dependent on the amino acids in these positions they have been called AVI, AAV, AAI, PAV, and PVI. The AVI and PAV haplotypes appear to be most frequent in Europe, Asia, and Africa. While in Europe and Asia the other haplotypes do not occur or are rare, Africans also frequently possess the AAI variant. In native Americans only the PAV variant was found. These haplotypes explained the bimodal distribution of PTC taste sensitivity in the Caucasian Utah sample and in a multi-racial population replica sample; the PAV haplotype was suggested to be the taster variant, whereas the AVI variant was suggested to be the non-taster variant (Kim et al. 2003b). The involvement of the *hTAS2R38* gene in PTC taste sensitivity has also been confirmed in a genetically isolated population from eastern Sardinia (Prodi et al. 2004). Functional expression studies in heterologous cells then showed that the five different *hTAS2R38* haplotypes encode operatively distinct receptors, the PAV variant being most sensitive to PTC and PROP and the AVI variant being insensitive, while the AAV, AAI, and PVI variants display similar intermediate sensitivities to these substances (Bufe et al. 2005). The taste responses of carriers of the PAV, AVI and AAI variants to PTC correlated strongly with the responses of the recombinant receptor variants. Thus, the hTAS2R38 polymorphisms code for functionally distinct receptor types that directly affect the bitterness of PTC and related N–C=S-containing compounds. Since different TAS2R alleles can code for functionally distinct bitter taste receptors, humans may not have only ~25 bitter taste receptors, but as many as there are TAS2R alleles (Bufe et al. 2005). The hTAS2R38 gene provides the first example that variations in a bitter taste receptor gene are the molecular basis for perceptual differences in the population. Similar genetic variations leading to missense mutations have been reported for the *hTAS2R4* and *hTAS2R5* genes (Ueda et al. 2001). Moreover, the human genome SNP database lists numerous SNPs that occur in the coding regions of *TAS2R* genes, and many occur in high frequencies. At least 17 hTAS2R genes are affected by at least one SNP leading to a missense codon, but most contain several such SNPs. Even nonsense mutations occur in the *hTAS2R44*, *hTAS2R45*, and *hTAS2R46* genes. So far, none of them has been analyzed functionally. As more TAS2Rs become deorphaned it will be possible to investigate the functional consequences of SNPs in TAS2Rs. Probably not all will result in functionally distinct bitter taste receptors, but several will. We may therefore expect that humans differ widely in their responses to bitter tastants with the implication that there will also be consequences for dietary habits and health.

Outlook

Clearly, since the identification and cloning of the first TAS2R family members (Adler et al. 2000; Matsunami et al. 2000), we have learned a lot about the molecular basis of bitter taste. Anatomical, genetic, functional, and behavioral evidence in mice and humans undoubtedly demonstrated that TAS2Rs mediate bitter taste. Analysis of the receptor repertoires in rodents, non-human primates, and humans revealed insights into the molecular evolution of this gene family in the context of hominid diets. We begin to understand the molecular architecture of the TAS2Rs and to recognize their agonist spectra. We also obtained the first impressions of how receptor gene variants determine inter-individual perceptual differences. And last but not least we learned about peripheral taste quality coding. Where is the research demand for the future? Several areas may be primary targets. Cracking the neural code of gustatory information should be at the top of the agenda. The power of knockout and transgene technology has proven extremely helpful. Possibly trans-synaptic markers targeted by these techniques to specific TRC could reveal how taste buds are connected to the afferent nerves and how these are connected to second and third order neurons and so on. The improvement of functional magnetic resonance tomography, and, in parallel, the understanding of the development of taste buds, their lineages, and the intercellular communication within the buds, will add to the understanding of the code. More practically, deorphaning of TAS2Rs, optimization of the receptor assays and better knowledge about the functional architecture of the TAS2Rs would be useful for the identification of compounds blocking the bitter tastes of medicines, thereby improving quality of life, particularly for children, the elderly, and chronic patients. In the long run this may also open the way to developing "artificial tongues" useful for the quantification of taste during food production processes and quality control. Another important research aim is to elucidate the extent to which TAS2R gene polymorphisms determine perceptual diversity. If so, does perceptual diversity in the population account for different diets and in this way affect health? Time will pass before the expected results and knowledge are obtained. But the cloning of the receptors allows researchers to develop the appropriate experimental tools and has thus programmed all future success.

Acknowledgements. The author thanks Dr. M. Behrens (Potsdam) for providing Fig. 2 and for helpful comments on the manuscript. The editorial assistance of Mrs. M. Lorse is also gratefully acknowledged. Original work carried out in the author's laboratory and referred to here was supported by a grant from the German Science Foundation (Me 1024/2–2).

References

Adler E, Hoon MA, Mueller KL, Chandrashekar J, Ryba NJ, Zuker CS (2000) A novel family of mammalian taste receptors [see comments]. Cell 100:693–702

Akabas MH, Dodd J, Al-Awqati Q (1988) A bitter substance induces a rise in intracellular calcium in a subpopulation of rat taste cells. Science 242:1047–1050

Ammon C, Schafer J, Kreuzer OJ, Meyerhof W (2002) Presence of a plasma membrane targeting sequence in the amino-terminal region of the rat somatostatin receptor 3. Arch Physiol Biochem 110:137–145

Andres-Barquin PJ, Conte C (2004) Molecular basis of bitter taste: the T2R family of G protein-coupled receptors. Cell Biochem Biophys 41:99–112

Aspen J, Gatch MB, Woods JH (1999) Training and characterization of a quinine taste discrimination in rhesus monkeys. Psychopharmacology 141:251–257

Barnicot NA, Harris H, Kalmus H (1951) Taste thresholds of further eighteen compounds and their correlation with P.T.C thresholds. Ann Eugen 16:119–128

Barratt-Fornell A, Drewnowski A (2002) The taste of health: nature's bitter gifts. Nutr Today 37:144–150
Bartoshuk LM (2000) Comparing sensory experiences across individuals: recent psychophysical advances illuminate genetic variation in taste perception. Chem Senses 25:447–460
Behrens M, Brockhoff A, Kuhn C, Bufe B, Winnig M, Meyerhof W (2004) The human taste receptor hTAS2R14 responds to a variety of different bitter compounds. Biochem Biophys Res Commun 319:479–485
Belitz H-D, Wieser H (1985) Bitter compounds: occurrence and structure-activity relationship. Food Rev Int 1:271–354
Biere A, Marak HB, Van Damme JM (2004) Plant chemical defense against herbivores and pathogens: generalized defense or trade-offs? Oecologia 140:430–441
Bockaert J, Pin JP (1999) Molecular tinkering of G protein-coupled receptors: an evolutionary success. EMBO J 18:1723–1729
Bufe B, Hofmann T, Krautwurst D, Raguse JD, Meyerhof W (2002) The human TAS2R16 receptor mediates bitter taste in response to beta-glucopyranosides. Nat Genet 32:397–401
Bufe B, Schöley-Pohl E, Krautwurst D, Hofmann T, Meyerhof W (2004) Identification of human bitter taste receptors. Am Chem Soc Symp S 867:45–59
Bufe B, Breslin PA, Kuhn C, Reed DR, Tharp CD, Slack JP, Kim UK, Drayna D, Meyerhof W (2005) The molecular basis of individual differences in phenylthiocarbamide and propylthiouracil bitterness perception. Curr Biol 15:322–327
Caicedo A, Roper SD (2001) Taste receptor cells that discriminate between bitter stimuli. Science 291:1557–1560
Caicedo A, Kim KN, Roper SD (2002) Individual mouse taste cells respond to multiple chemical stimuli. J Physiol 544:501–509
Caicedo A, Pereira E, Margolskee RF, Roper SD (2003) Role of the G-protein subunit alpha-gustducin in taste cell responses to bitter stimuli. J Neurosci 23:9947–9952
Chan CY, Yoo JE, Travers SP (2004) Diverse bitter stimuli elicit highly similar patterns of fos-like immunoreactivity in the nucleus of the solitary tract. Chem Senses 29:573–581
Chandrashekar J, Mueller KL, Hoon MA, Adler E, Feng L, Guo W, Zuker CS, Ryba NJ (2000) T2Rs function as bitter taste receptors. Cell 100:703–711
Clapham DE (2003) TRP channels as cellular sensors. Nature 426:517–524
Clapp TR, Stone LM, Margolskee RF, Kinnamon SC (2001) Immunocytochemical evidence for co-expression of type III IP3 receptor with signaling components of bitter taste transduction. BMC Neurosci 2:6
Clapp TR, Yang R, Stoick CL, Kinnamon SC, Kinnamon JC (2004) Morphologic characterization of rat taste receptor cells that express components of the phospholipase C signaling pathway. J Comp Neurol 468:311–321
Conte C, Ebeling M, Marcuz A, Nef P, Andres-Barquin PJ (2002) Identification and characterization of human taste receptor genes belonging to the TAS2R family. Cytogenet Genome Res 98:45–53
Conte C, Ebeling M, Marcuz A, Nef P, Andres-Barquin PJ (2003) Evolutionary relationships of the Tas2r receptor gene families in mouse and human. Physiol Genomics 14:73–82
Czepa A, Hofmann T (2003) Structural and sensory characterization of compounds contributing to the bitter off-taste of carrots (Daucus carota L.) and carrot puree. J Agric Food Chem 51:3865–3873
Czepa A, Hofmann T (2004) Quantitative studies and sensory analyses on the influence of cultivar, spatial tissue distribution, and industrial processing on the bitter off-taste of carrots (Daucus carota l.) and carrot products. J Agric Food Chem 52:4508–4514
Dahl M, Erickson RP, Simon SA (1997) Neural responses to bitter compounds in rats. Brain Res 756:22–34
Damak S, Rong M, Yasumatsu K, Kokrashvili Z, Varadarajan V, Zou S, Jiang P, Ninomiya Y, Margolskee RF (2003) Detection of sweet and umami taste in the absence of taste receptor T1r3. Science 301:850–853
Danilova V, Hellekant G (2003) Comparison of the responses of the chorda tympani and glossopharyngeal nerves to taste stimuli in C57BL/6J mice. BMC Neurosci 4:5
Danilova V, Hellekant G (2004) Sense of taste in a New World monkey, the common marmoset. II. Link between behavior and nerve activity. J Neurophysiol 92:1067–1076
Denton D (1982) The hunger for salt; an anthropological and medical analysis. Springer, Berlin Heidelberg New York
Di Lorenzo PM, Lemon CH (2000) The neural code for taste in the nucleus of the solitary tract of the rat: effects of adaptation. Brain Res 852:383–397
Drayna D, Coon H, Kim UK, Elsner T, Cromer K, Otterud B, Baird L, Peiffer AP, Leppert M (2003) Genetic analysis of a complex trait in the Utah Genetic Reference Project: a major locus for PTC taste ability on chromosome 7q and a secondary locus on chromosome 16p. Hum Genet 6:6
Drewnowski A, Gomez-Carneros C (2000) Bitter taste, phytonutrients, and the consumer: a review [in process citation]. Am J Clin Nutr 72:1424–1435

Duncan LM, Deeds J, Hunter J, Shao J, Holmgren LM, Woolf EA, Tepper RI, Shyjan AW (1998) Down-regulation of the novel gene melastatin correlates with potential for melanoma metastasis. Cancer Res 58:1515–1520

Engel E, Septier C, Leconte N, Salles C, Le Quere JL (2001) Determination of taste-active compounds of a bitter Camembert cheese by omission tests. J Dairy Res 68:675–688

Finger TE, Bottger B, Hansen A, Anderson KT, Alimohammadi H, Silver WL (2003) Solitary chemoreceptor cells in the nasal cavity serve as sentinels of respiration. Proc Natl Acad Sci USA 100:8981–8986

Fischer A, Gilad Y, Man O, Paabo S (2005) Evolution of bitter taste receptors in humans and apes. Mol Biol Evol 22:432–436

Frank O, Hofmann T (2002) Reinvestigation of the chemical structure of bitter-tasting quinizolate and homo-quinizolate and studies on their maillard-type formation pathways using suitable (13)c-labeling experiments. J Agric Food Chem 50:6027–6036

Ganshirt H (1953) Isolation of aristolochic acid from various Aristolochiaceae and its quantitative determination. Pharmazie 8:584–592

Gienapp E, Schröder KL (1975) Struktur und Bittere in der Huluponreihe. Nahrung 19:697–706

Gilbertson TA, Boughter JD Jr, Zhang H, Smith DV (2001) Distribution of gustatory sensitivities in rat taste cells: whole-cell responses to apical chemical stimulation. J Neurosci 21:4931–4941

Go Y, Satta Y, Takenaka O, Takahata N (2005) Lineage-specific loss of function of bitter taste receptor genes in humans and non-human primates. Genetics http://dx.doi.org/10.1534/genetics.104.037523

Goren-Inbar N, Alperson N, Kislev ME, Simchoni O, Melamed Y, Ben-Nun A, Werker E (2004) Evidence of hominin control of fire at Gesher Benot Ya'aqov, Israel. Science 304:725–727

Guo SW, Reed DR (2001) The genetics of phenylthiocarbamide perception. Ann Hum Biol 28:111–142

Gutierrez-Rosales F, Rios JJ, Gomez-Rey ML (2003) Main polyphenols in the bitter taste of virgin olive oil. Structural confirmation by on-line high-performance liquid chromatography electrospray ionization mass spectrometry. J Agric Food Chem 51:6021–6025

Habibi-Najafi MB, Lee BH (1996) Bitterness in cheese: a review. Crit Rev Food Sci Nutr 36:397–411

Hamor GH (1961) Correlation of chemical structure and taste in the saccharin series. Science 131:1416–1417

Harris H, Kalmus H (1949) Chemical specificity in genetical differences of taste sensitivity. Ann Eugen 15:32–45

He W, Danilova V, Zou S, Hellekant G, Max M, Margolskee RF, Damak S (2002) Partial rescue of taste responses of alpha-gustducin null mice by transgenic expression of alpha-transducin. Chem Senses 27:719–727

Hellekant G, Ninomiya Y, Danilova V (1998) Taste in chimpanzees. III. Labeled-line coding in sweet taste. Physiol Behav 65:191–200

Hilliard MA, Bergamasco C, Arbucci S, Plasterk RH, Bazzicalupo P (2004) Worms taste bitter: ASH neurons, QUI-1, GPA-3 and ODR-3 mediate quinine avoidance in *Caenorhabditis elegans*. EMBO J 23:1101–1111

Höfer D, Puschel B, Drenckhahn D (1996) Taste receptor-like cells in the rat gut identified by expression of alpha-gustducin. Proc Natl Acad Sci USA 93:6631–6634

Hofmann T, Chubanov V, Gudermann T, Montell C (2003) TRPM5 is a voltage-modulated and Ca(2+)-activated monovalent selective cation channel. Curr Biol 13:1153–1158

Horne J, Lawless HT, Speirs W, Sposato D (2002) Bitter taste of saccharin and acesulfame-K. Chem Senses 27:31–38

Huang L, Shanker YG, Dubauskaite J, Zheng JZ, Yan W, Rosenzweig S, Spielman AI, Max M, Margolskee RF (1999) Ggamma13 colocalizes with gustducin in taste receptor cells and mediates IP3 responses to bitter denatonium. Nat Neurosci 2:1055–1062

Huang YJ, Maruyama Y, Lu KS, Pereira E, Plonsky I, Baur JE, Wu D, Roper SD (2005) Mouse taste buds use serotonin as a neurotransmitter. J Neurosci 25:843–847

Hwang PM, Verma A, Bredt DS, Snyder SH (1990) Localization of phosphatidylinositol signaling components in rat taste cells: role in bitter taste transduction. Proc Natl Acad Sci USA 87:7395–7399

Jiang P, Cui M, Zhao B, Liu Z, Snyder LA, Benard LM, Osman R, Margolskee RF, Max M (2005) Lactisole interacts with the transmembrane domains of human T1R3 to inhibit sweet taste. J Biol Chem 280:15238–15246

Katz A, Wu D, Simon MI (1992) Subunits beta gamma of heterotrimeric G protein activate beta 2 isoform of phospholipase C. Nature 360:686–689

Katz DB, Nicolelis MA, Simon SA (2002) Gustatory processing is dynamic and distributed. Curr Opin Neurobiol 12:448–454

Keast RS, Breslin PA (2002) Cross-adaptation and bitterness inhibition of L-tryptophan, L-phenylalanine and urea: further support for shared peripheral physiology. Chem Senses 27:123–131

Kim MR, Kusakabe Y, Miura H, Shindo Y, Ninomiya Y, Hino A (2003a) Regional expression patterns of taste receptors and gustducin in the mouse tongue. Biochem Biophys Res Commun 312:500–506

Kim UK, Jorgenson E, Coon H, Leppert M, Risch N, Drayna D (2003b) Positional cloning of the human quantitative trait locus underlying taste sensitivity to phenylthiocarbamide. Science 299:1221–1225

Kingsbury J (1964) Poisonous plants of the United States and Canada. Prentice Hall, Englewood Cliffs

Kock K, Morley SD, Mullins JJ, Schmale H (1994) Denatonium bitter tasting among transgenic mice expressing rat von Ebner's gland protein. Physiol Behav 56:1173–1177

Kolesnikov SS, Margolskee RF (1995) A cyclic-nucleotide-suppressible conductance activated by transducin in taste cells. Nature 376:85–88

Krautwurst D, Yau KW, Reed RR (1998) Identification of ligands for olfactory receptors by functional expression of a receptor library. Cell 95:917–926

Kretz O, Barbry P, Bock R, Lindemann B (1999) Differential expression of RNA and protein of the three pore-forming subunits of the amiloride-sensitive epithelial sodium channel in taste buds of the rat. J Histochem Cytochem 47:51–64

Kubo I (1994) Structural basis for bitterness based on Rabdosia diterpenes. Physiol Behav 56:1203–1207

Kuhn C, Bufe B, Winnig M, Hofmann T, Frank O, Behrens M, Lewtschenko T, Slack JP, Ward CD, Meyerhof W (2004) Bitter taste receptors for saccharin and acesulfame K. J Neurosci 24:10260–10265

Lechtenberg M, Nahrstedt A (1999) Cyanogenic glycosides. In: Ikan R (ed) Naturally occurring glycosides. Wiley, Chichester, pp 147–191

Lesschaeve I, Noble AC (2005) Polyphenols: factors influencing their sensory properties and their effects on food and beverage preferences. Am J Clin Nutr 81:330S–335S

Li X, Staszewski L, Xu H, Durick K, Zoller M, Adler E (2002) Human receptors for sweet and umami taste. Proc Natl Acad Sci USA 99:4692–4696

Lindemann B (1996) Taste reception. Physiol Rev 76:718–766

Lindemann B (1999) Receptor seeks ligand: on the way to cloning the molecular receptors for sweet and bitter taste. Nat Med 5:381–382 [news]

Lindemann B (2001) Receptors and transduction in taste. Nature 413:219–225

Liu D, Liman ER (2003) Intracellular Ca^{2+} and the phospholipid PIP2 regulate the taste transduction ion channel TRPM5. Proc Natl Acad Sci USA 100:15160–15165

Lundy RF Jr, Contreras RJ (1999) Gustatory neuron types in rat geniculate ganglion. J Neurophysiol 82:2970–2988

Lush IE (1981) The genetics of tasting in mice. I. Sucrose octaacetate. Genet Res 38:93–95

Lush IE (1984) The genetics of tasting in mice. III. Quinine. Genet Res 44:151–160

Lush IE (1986) The genetics of tasting in mice. IV. The acetates of raffinose, galactose and beta-lactose. Genet Res 47:117–123

Lush IE, Holland G (1988) The genetics of tasting in mice. V. Glycine and cycloheximide. Genet Res 52:207–212

Martin JH (1989) Neuroanatomy. Prentice Hall International, London

Matsumoto A, Kawaguchi N, Asaka Y, Kubota T (1986) Relationship between lipophilicity, bitterness and structure of phenyl beta-D-glucopyranosides. Food Chem 21:83–92

Matsunami H, Amrein H (2003) Taste and pheromone perception in mammals and flies. Genome Biol 4:220

Matsunami H, Montmayeur JP, Buck LB (2000) A family of candidate taste receptors in human and mouse. Nature 404:601–604 [see comments]

Matsuo R (2000) Role of saliva in the maintenance of taste sensitivity. Crit Rev Oral Biol Med 11:216–229

Max M, Shanker YG, Huang L, Rong M, Liu Z, Campagne F, Weinstein H, Damak S, Margolskee RF (2001) Tas1r3, encoding a new candidate taste receptor, is allelic to the sweet responsiveness locus Sac. Nat Genet 28:58–63

McGregor R (2004) Taste modification in the biotech era. Food Technol 58:24–30

McLaughlin SK, McKinnon PJ, Margolskee RF (1992) Gustducin is a taste-cell-specific G protein closely related to the transducins. Nature 357:563–569

McLaughlin SK, McKinnon PJ, Spickofsky N, Danho W, Margolskee RF (1994) Molecular cloning of G proteins and phosphodiesterases from rat taste cells. Physiol Behav 56:1157–1164

Medler KF, Margolskee RF, Kinnamon SC (2003) Electrophysiological characterization of voltage-gated currents in defined taste cell types of mice. J Neurosci 23:2608–2617

Meyerhof W, Kuhn C, Brockhoff A, Winnig M, Bufe B, Schöley-Pohl E, Behrens M (2005) Peripheral receptors and transduction mechanisms in taste perception. In: Hofmann T, Rothe M, Schieberle P (eds) Proceedings of the 8th Wartburg Symposium. State-of-the-art in flavour chemistry and biology. Deutsche Forschungsanstalt für Lebensmittelchemie, Garching

Miller IJ Jr (1995) Anatomy of the peripheral taste system. In: Doty RL (ed) Handbook of olfaction and gustation. Dekker, New York, pp 521–547

Milton K (2003) The critical role played by animal source foods in human (Homo) evolution. J Nutr 133:3886S–3892S

Ming D, Ruiz-Avila L, Margolskee RF (1998) Characterization and solubilization of bitter-responsive receptors that couple to gustducin. Proc Natl Acad Sci USA 95:8933–8938

Ming D, Ninomiya Y, Margolskee RF (1999) Blocking taste receptor activation of gustducin inhibits gustatory responses to bitter compounds. Proc Natl Acad Sci USA 96:9903–9908

Misaka T, Kusakabe Y, Emori Y, Gonoi T, Arai S, Abe K (1997) Taste buds have a cyclic nucleotide-activated channel, CNGgust. J Biol Chem 272:22623–22629

Miwa K, Kanemura F, Tonosaki K (1997) Tastes activate different second messengers in taste cells. J Vet Med Sci 59:81–83

Miyoshi MA, Abe K, Emori Y (2001) IP(3) receptor type 3 and PLCbeta2 are co-expressed with taste receptors T1R and T2R in rat taste bud cells. Chem Senses 26:259–265

Mombaerts P (2004) Genes and ligands for odorant, vomeronasal and taste receptors. Nat Rev Neurosci 5:263–278

Montmayeur JP, Liberles SD, Matsunami H, Buck LB (2001) A candidate taste receptor gene near a sweet taste locus. Nat Neurosci 4:492–498

Moriyama M, Oba K (2004) Sprouts as antioxidant food resources and young people's taste for them. Biofactors 21:247–249

Mueller KL, Hoon MA, Erlenbach I, Chandrshekar J, Zuker CS, Ryba NJP (2005) The receptors and coding logic for bitter taste. Nature 434:225–229

Murata Y, Sata NU (2000) Isolation and structure of pulcherrimine, a novel bitter-tasting amino acid, from the sea urchin (*Hemicentrotus pulcherrimus*) ovaries. J Agric Food Chem 48:5557–5560

Nakashima K, Ninomiya Y (1998) Increase in inositol 1,4,5-triphosphate levels of the fungiform papilla in response to saccharin and bitter substances in mice. Cell Physiol Biochem 8:224–230

Nelson G, Hoon MA, Chandrashekar J, Zhang Y, Ryba NJ, Zuker CS (2001) Mammalian sweet taste receptors. Cell 106:381–390

Nelson G, Chandrashekar J, Hoon MA, Feng L, Zhao G, Ryba NJ, Zuker CS (2002) An amino-acid taste receptor. Nature 416:199–202

Nishijo H, Norgren R (1997) Parabrachial neural coding of taste stimuli in awake rats. J Neurophysiol 78:2254–2268

Nolte DL, Mason JR, Lewis SL (1994) Tolerance of bitter compounds by an herbivore, *Cavia porcellus*. J Chem Ecol 20:303–308

Northcutt RG (2004) Taste buds: development and evolution. Brain Behav Evol 64:198–206

Offermanns S, Simon MI (1995) G alpha 15 and G alpha 16 couple a wide variety of receptors to phospholipase C. J Biol Chem 270:15175–15180

Ogura T, Mackay-Sim A, Kinnamon SC (1997) Bitter taste transduction of denatonium in the mudpuppy *Necturus maculosus*. J Neurosci 17:3580–3587

Ozeck M, Brust P, Xu H, Servant G (2004) Receptors for bitter, sweet and umami taste couple to inhibitory G protein signaling pathways. Eur J Pharmacol 489:139–149

Parry CM, Erkner A, le Coutre J (2004) Divergence of T2R chemosensory receptor families in humans, bonobos, and chimpanzees. Proc Natl Acad Sci USA 101:14830–14834

Perez CA, Huang L, Rong M, Kozak JA, Preuss AK, Zhang H, Max M, Margolskee RF (2002) A transient receptor potential channel expressed in taste receptor cells. Nat Neurosci 5:1169–1176

Prawitt D, Monteilh-Zoller MK, Brixel L, Spangenberg C, Zabel B, Fleig A, Penner R (2003) TRPM5 is a transient Ca^{2+}-activated cation channel responding to rapid changes in $[Ca^{2+}]_i$. Proc Natl Acad Sci USA 100:15166–15171

Prodi DA, Drayna D, Forabosco P, Palmas MA, Maestrale GB, Piras D, Pirastu M, Angius A (2004) Bitter taste study in a Sardinian genetic isolate supports the association of phenylthiocarbamide sensitivity to the TAS2R38 bitter receptor gene. Chem Senses 29:697–702

Pronin AN, Tang H, Connor J, Keung W (2004) Identification of ligands for two human bitter T2R receptors. Chem Senses 29:583–593

Raksakulthai R, Haard NF (2003) Exopeptidases and their application to reduce bitterness in food: a review. Crit Rev Food Sci Nutr 43:401–445

Redfern CH, Coward P, Degtyarev MY, Lee EK, Kwa AT, Hennighausen L, Bujard H, Fishman GI, Conklin BR (1999) Conditional expression and signaling of a specifically designed Gi-coupled receptor in transgenic mice. Nat Biotechnol 17:165–169

Reed DR, Nanthakumar E, North M, Bell C, Bartoshuk LM, Price RA (1999) Localization of a gene for bitter-taste perception to human chromosome 5p15. Am J Hum Genet 64:1478–1480

Rolls ET (1995) Central taste anatomy and neurophysiology. In: Doty RL (eds) Handbook of olfaction and gustation. Dekker, New York, pp 549–573

Rolls ET (2004) The functions of the orbitofrontal cortex. Brain Cogn 55:11–29

Roper SD (1992) The microphysiology of peripheral taste organs. J Neurosci 12:1127–1134

Rössler P, Kroner C, Freitag J, Noe J, Breer H (1998) Identification of a phospholipase C beta subtype in rat taste cells. Eur J Cell Biol 77:253–261

Rossler P, Boekhoff I, Tareilus E, Beck S, Breer H, Freitag J (2000) G protein betagamma complexes in circumvallate taste cells involved in bitter transduction. Chem Senses 25:413–421

Rouseff RL (1980) Flavonoids and citrus quality. ACS Symposium Series 143:83–108

Rozin P, Vollmecke TA (1986) Food likes and dislikes. Annu Rev Nutr 6:433–456

Ruiz-Avila L, McLaughlin SK, Wildman D, McKinnon PJ, Robichon A, Spickofsky N, Margolskee RF (1995) Coupling of bitter receptor to phosphodiesterase through transducin in taste receptor cells. Nature 376:80–85

Ruiz-Avila L, Ming D, Margolskee RF (2000) An in vitro assay useful to determine the potency of several bitter compounds. Chem Senses 25:361–368

Saito H, Kubota M, Roberts RW, Chi Q, Matsunami H (2004) RTP family members induce functional expression of mammalian odorant receptors. Cell 119:679–691

Saper CB, Chou TC, Elmquist JK (2002) The need to feed: homeostatic and hedonic control of eating. Neuron 36:199–211

Saroli A (1984) Structure-activity relationship of a bitter compound: denatonium chloride. Naturwissenschaften 71:428–429

Sato T, Beidler LM (1997) Broad tuning of rat taste cells for four basic taste stimuli. Chem Senses 22:287–293

Sbarbati A, Osculati F (2003) Solitary chemosensory cells in mammals? Cells Tissues Organs 175:51–55

Sbarbati A, Crescimanno C, Bernardi P, Osculati F (1999) Alpha-gustducin-immunoreactive solitary chemosensory cells in the developing chemoreceptorial epithelium of the rat vallate papilla. Chem Senses 24:469–472

Sbarbati A, Merigo F, Benati D, Tizzano M, Bernardi P, Osculati F (2004) Laryngeal chemosensory clusters. Chem Senses 29:683–692

Schieberle P, Hofmann T (2003) Die molekulare Welt des Lebensmittelgenusses. Chem Unserer Zeit 37:388–401

Schiffman SS (2000) Taste quality and neural coding: implications from psychophysics and neurophysiology. Physiol Behav 69:147–159

Schiffman SS, Gatlin CA (1993) Sweeteners: state of knowledge review. Neurosci Biobehav Rev 17:313–345

Schiffman SS, Zervakis J, Heffron S, Heald AE (1999) Effect of protease inhibitors on the sense of taste. Nutrition 15:767–772

Schmale H, Bamberger C (1997) A novel protein with strong homology to the tumor suppressor p53. Oncogene 15:1363–1367

Schmale H, Holtgreve-Grez H, Christiansen H (1990) Possible role for salivary gland protein in taste reception indicated by homology to lipophilic-ligand carrier proteins. Nature 343:366–369

Schoenfeld MA, Neuer G, Tempelmann C, Schussler K, Noesselt T, Hopf JM, Heinze HJ (2004) Functional magnetic resonance tomography correlates of taste perception in the human primary taste cortex. Neuroscience 127:347–353

Scott TR, Giza BK (2000) Issues of gustatory neural coding: where they stand today. Physiol Behav 69:65–76

Scott TR, Giza BK, Yan J (1999) Gustatory neural coding in the cortex of the alert cynomolgus macaque: the quality of bitterness. J Neurophysiol 81:60–71

Serafini M, Bugianesi R, Maiani G, Valtuena S, De Santis S, Crozier A (2003) Plasma antioxidants from chocolate. Nature 424:1013

Shallenberger RS, Acree TE (1967) Molecular theory of sweet taste. Nature 216:480–482

Shi P, Zhang J, Yang H, Zhang YP (2003) Adaptive diversification of bitter taste receptor genes in Mammalian evolution. Mol Biol Evol 20:805–814

Shirokova E, Schmiedeberg K, Bedner P, Niessen H, Willecke K, Raguse JD, Meyerhof W, Krautwurst D (2005) Identification of specific ligands for orphan olfactory receptors: G protein-dependent agonism and antagonism of odorants. J Biol Chem 280:11807–11815

Smith DV, Frank ME (1993) Sensory coding by peripheral taste fibers. In: Simon SA, Frank ME (eds) Mechanisms of taste transduction. CRC, Boca Raton, pp 295–338

Smith DV, St John SJ (1999) Neural coding of gustatory information. Curr Opin Neurobiol 9:427–435

Spector AC, Kopka SL (2002) Rats fail to discriminate quinine from denatonium: implications for the neural coding of bitter-tasting compounds. J Neurosci 22:1937–1941

Spielman AI, Huque T, Nagai H, Whitney G, Brand JG (1994) Generation of inositol phosphates in bitter taste transduction. Physiol Behav 56:1149–1155

Spielman AI, Nagai H, Sunavala G, Dasso M, Breer H, Boekhoff I, Huque T, Whitney G, Brand JG (1996) Rapid kinetics of second messenger production in bitter taste. Am J Physiol 270:C926–C931

Steiner JE (1994) Behavior manifestations indicative of hedonics and intensity in chemosensory experience. In: Kurihara K, Suzuki N, Ogawa H (eds) Olfaction and taste XI. Springer, Tokyo, pp 284–287

Stevens DR, Seifert R, Bufe B, Muller F, Kremmer E, Gauss R, Meyerhof W, Kaupp UB, Lindemann B (2001) Hyperpolarization-activated channels HCN1 and HCN4 mediate responses to sour stimuli. Nature 413:631–635

Straub SG, Mulvaney-Musa J, Yajima H, Weiland GA, Sharp GW (2003) Stimulation of insulin secretion by denatonium, one of the most bitter-tasting substances known. Diabetes 52:356–364

Sugino Y, Umemoto A, Mizutani S (2002) Insensitivity to the bitter taste of chloramphenicol: an autosomal recessive trait. Genes Genet Syst 77:59–62

Suzuki H, Onishi H, Takahashi Y, Iwata M, Machida Y (2003) Development of oral acetaminophen chewable tablets with inhibited bitter taste. Int J Pharm 251:123–132

Tabata S, Wada A, Kobayashi T, Nishimura S, Muguruma M, Iwamoto H (2003) Bovine circumvallate taste buds: taste cell structure and immunoreactivity to alpha-gustducin. Anat Rec 271A:217–224

Takami S, Getchell TV, McLaughlin SK, Margolskee RF, Getchell ML (1994) Human taste cells express the G protein alpha-gustducin and neuron-specific enolase. Brain Res Mol Brain Res 22:193–203

Tancredi T, Lelj F, Temussi PA (1979) Three-dimensional mapping of the bitter taste receptor site. Chem Senses Flavour 4:259–265

Tepper BJ (1998) 6-n-Propylthiouracil: a genetic marker for taste, with implications for food preference and dietary habits. Am J Hum Genet 63:1271–1276

Tobach E, Bellin JS, Das DK (1974) Differences in bitter taste perception in three strains of rats. Behav Genet 4:405–410

Travers SP (2002) Quinine and citric acid elicit distinctive Fos-like immunoreactivity in the rat nucleus of the solitary tract. Am J Physiol Regul Integr Comp Physiol 282:R1798–R1810

Uchida T, Tanigake A, Miyanaga Y, Matsuyama K, Kunitomo M, Kobayashi Y, Ikezaki H, Taniguchi A (2003) Evaluation of the bitterness of antibiotics using a taste sensor. J Pharm Pharmacol 55:1479–1485

Ueda T, Ugawa S, Ishida Y, Shibata Y, Murakami S, Shimada S (2001) Identification of coding single-nucleotide polymorphisms in human taste receptor genes involving bitter tasting. Biochem Biophys Res Commun 285:147–151

Ueda T, Ugawa S, Yamamura H, Imaizumi Y, Shimada S (2003) Functional interaction between T2R taste receptors and G-protein {alpha} subunits expressed in taste receptor cells. J Neurosci 23:7376–7380

Verhagen JV, Giza BK, Scott TR (2003) Responses to taste stimulation in the ventroposteromedial nucleus of the thalamus in rats. J Neurophysiol 89:265–275

von Skramlik E (1926) Handbuch der Physiologie der niederen Sinne. Thieme, Leipzig

Wang X, Thomas SD, Zhang J (2004) Relaxation of selective constraint and loss of function in the evolution of human bitter taste receptor genes. Hum Mol Genet 13:2671–2678

Wetzel CH, Oles M, Wellerdieck C, Kuczkowiak M, Gisselmann G, Hatt H (1999) Specificity and sensitivity of a human olfactory receptor functionally expressed in human embryonic kidney 293 cells and *Xenopus laevis* oocytes. J Neurosci 19:7426–7433

Whitehead MC, Ganchrow JR, Ganchrow D, Yao B (1999) Organization of geniculate and trigeminal ganglion cells innervating single fungiform taste papillae: a study with tetramethylrhodamine dextran amine labeling. Neuroscience 93:931–941

Wong GT, Gannon KS, Margolskee RF (1996a) Transduction of bitter and sweet taste by gustducin. Nature 381:796–800

Wong GT, Ruiz-Avila L, Ming D, Gannon KS, Margolskee RF (1996b) Biochemical and transgenic analysis of gustducin's role in bitter and sweet transduction. Cold Spring Harb Symp Quant Biol 61:173–184

Wong GT, Ruiz-Avila L, Margolskee RF (1999) Directing gene expression to gustducin-positive taste receptor cells. J Neurosci 19:5802–5809

Wooding S, Kim UK, Bamshad MJ, Larsen J, Jorde LB, Drayna D (2004) Natural selection and molecular evolution in PTC, a bitter-taste receptor gene. Am J Hum Genet 74:637–646

Wu SV, Rozengurt N, Yang M, Young SH, Sinnett-Smith J, Rozengurt E (2002) Expression of bitter taste receptors of the T2R family in the gastrointestinal tract and enteroendocrine STC-1 cells. Proc Natl Acad Sci USA 99:2392–2397

Yan W, Sunavala G, Rosenzweig S, Dasso M, Brand JG, Spielman AI (2001) Bitter taste transduced by PLC-beta(2)-dependent rise in IP(3) and alpha-gustducin-dependent fall in cyclic nucleotides. Am J Physiol Cell Physiol 280:C742–C751

Yang H, Wanner IB, Roper SD, Chaudhari N (1999) An optimized method for in situ hybridization with signal amplification that allows the detection of rare mRNAs. J Histochem Cytochem 47:431–446

Zagrobelny M, Bak S, Rasmussen AV, Jorgensen B, Naumann CM, Lindberg Moller B (2004) Cyanogenic glucosides and plant-insect interactions. Phytochemistry 65:293–306

Zhang Y, Hoon MA, Chandrashekar J, Mueller KL, Cook B, Wu D, Zuker CS, Ryba NJ (2003) Coding of sweet, bitter, and umami tastes. Different receptor cells sharing similar signaling pathways. Cell 112:293–301

Zhao GQ, Zhang Y, Hoon MA, Chandrashekar J, Erlenbach I, Ryba NJ, Zuker CS (2003) The receptors for mammalian sweet and umami taste. Cell 115:255–266

Rev Physiol Biochem Pharmacol (2005) 154:73–100
DOI 10.1007/s10254-005-0043-y

Hanns Ulrich Zeilhofer

Synaptic modulation in pain pathways

Published online: 30 July 2005
© Springer-Verlag 2005

Abstract All higher organisms possess a sensory system that allows them to detect potentially tissue-damaging (or noxious) stimuli. The proper functioning of this system is essential to protect their bodies from tissue damage. However, under pathological conditions after severe tissue injury and in inflammatory or neuropathic diseases, this system can become sensitized, and pain can then turn into a disease. Such exaggerated pain sensation (or hyperalgesia) can arise at different levels of integration. It can originate from an increased responsiveness of primary nociceptors, specialized nerve cells, which sense noxious stimuli, or from changes in the central processing of nociceptive input. Like other sensory input, nociceptive signals are relayed in the central nervous system by neurons, which communicate with each other mainly through chemical synapses. Changes in the excitability of these neurons or in the strength of their synaptic coupling provide the cellular basis for many forms of pathological pain. This review focuses on the synaptic processing of pain-related signals in the spinal cord dorsal horn, the first site of synaptic integration in the pain pathway. Particular emphasis is paid to synaptic processes underlying the generation of pathological pain evoked by inflammation or neuropathic diseases.

Introduction

Our bodies sense potentially tissue-damaging stimuli using specialized nerve cells called nociceptors. These are thinly myelinated or unmyelinated slowly conducting C and Aδ nerve fibers, which connect the peripheral tissue with the central nervous system (CNS). Whenever a noxious stimulus hits such a nociceptor, a generator potential is elicited, which, if it exceeds a certain threshold, evokes an action potential that travels along the peripheral nerve to the spinal cord dorsal horn. Here, the nociceptors make synaptic connections with projection neurons and local excitatory and inhibitory interneurons. Projection neurons transmit

Hanns Ulrich Zeilhofer (✉)
Universität Zürich, Institut für Pharmakologie und Toxikologie,
Winterthurerstrasse 190, 8057 Zürich, Switzerland
e-mail: Zeilhofer@pharma.unizh.ch · Tel.: +41-44-6355912 · Fax: +41-44-6359708

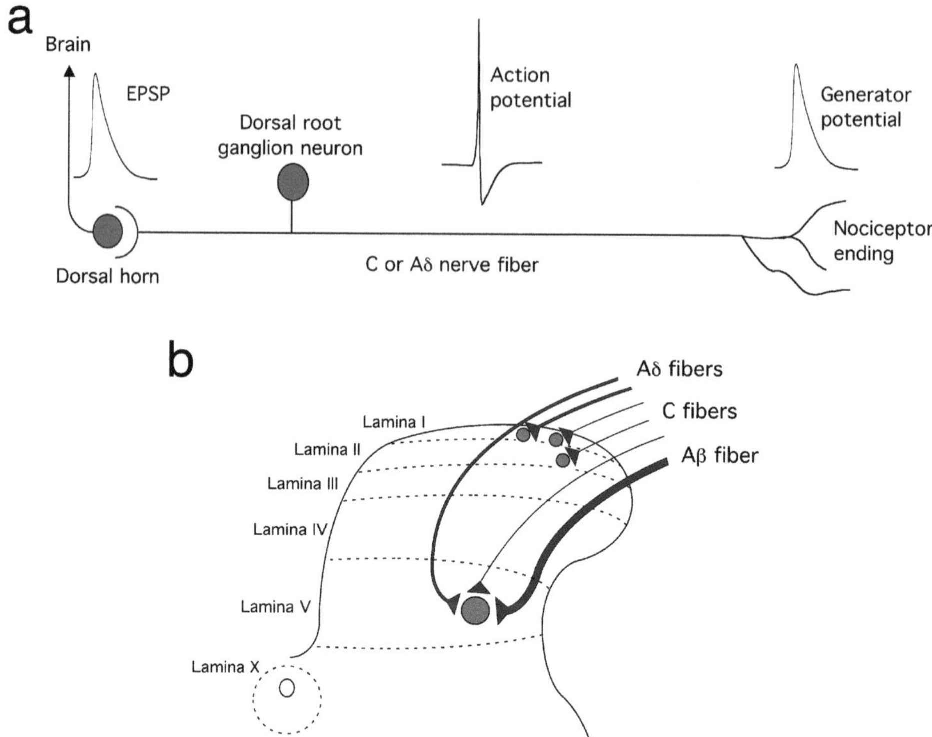

Fig. 1 Transmission of nociceptive signals from the peripheral tissue to the spinal cord dorsal horn. **a** Noxious stimuli elicit depolarizing generator potentials in the peripheral terminals of nociceptive C and Aδ fibers. Suprathreshold depolarizations evoke action potentials, which travel along the peripheral nerve to the spinal cord dorsal horn, where they make synaptic contact with projection neurons and excitatory and inhibitory interneurons. **b** Innervation of the dorsal horn by primary afferent nerve fibers. C fibers terminate mainly in laminae I, II, and V; Aδ fibers in laminae I and V; and Aβ fibers in the deep dorsal horn (laminae III, IV, and V).

the nociceptive information through several relay stations to higher brain areas, where the conscious phenomenon of pain arises (Fig. 1a).

The proper functioning of this nociceptive system is essential to protect the body from tissue damage. However, under pathological conditions, such as in inflammatory or neuropathic diseases, this system can become sensitized, and pain can then turn into a disease. Such exaggerated pain sensations can either originate from an increased responsiveness of peripheral nociceptors (primary or peripheral hyperalgesia) or from changes in the central processing of nociceptive input (secondary or central hyperalgesia). Central changes in nociceptive processing can also lead to allodynia, the phenomenon that stimuli not normally sensed as painful evoke pain.

The spinal cord dorsal horn is the first site of synaptic integration in the pain pathway and critically contributes to pathologically exaggerated pain sensations. According to Rexed, the spinal cord can be divided into 10 different laminae (I–X). Primary nociceptive afferents terminate mainly in the superficial dorsal horn in laminae I and II; the latter is also called substantia gelatinosa. A subset of primary nociceptive afferents also makes synaptic connections with so-called wide dynamic range neurons located in the deep dorsal horn, mostly in lamina V. In addition, the area around the central canal (lamina X) participates in the processing of nociceptive signals (Fig. 1b).

Synaptic transmission between primary afferent nerve fibers and dorsal horn projection neurons is not a fixed process but is subject to the dynamic control by local interneurons, descending pro- or antinociceptive pathways and chemical mediators released from neurons and glial cells.

This review will concentrate on the synaptic processing of nociceptive input in the spinal cord dorsal horn with particular emphasis on the contribution of specific isoforms of ion channels and receptors to inflammatory and neuropathic pain and their role during activity-dependent changes in spinal nociceptive processing.

Glutamatergic synaptic transmission in the spinal cord dorsal horn and its contribution to activity-dependent changes in nociceptive processing

AMPA receptors

Primary sensory neurons use the amino acid L-glutamate as their principle fast excitatory neurotransmitter. Synaptically released L-glutamate primarily acts on postsynaptic ionotropic glutamate receptors of the (±)-α-amino-3-hydroxy-5-methylisoxazole-4-propionic acid (AMPA), kainate, and N-methyl-D-aspartate (NMDA) subtypes and on G-protein-coupled (metabotropic) glutamate receptors. Four different types of AMPA receptor subunits, termed GluR-A through GluR-D or GluR-1 through GluR-4, are known. Although all four types are expressed in the spinal cord, their relative abundance varies considerably among the different laminae. GluR-A is most prevalent in laminae I and II, while GluR-B is rather homogeneously distributed throughout the dorsal horn, and GluR-C and GluR-D are relatively weakly expressed in the superficial laminae (Nagy et al. 2004). The strong expression in the superficial dorsal horn of GluR-A suggests a prominent role in spinal nociceptive processing. In the hippocampus, GluR-A plays an important role in activity-dependent plasticity, namely in long-term potentiation (LTP) at CA3-CA1 pyramidal cell synapses. Induction of LTP increases the phosphorylation of GluR-A by calcium- and calmodulin-dependent kinase II (CaMKII) at Ser 831 (Barria et al. 1997). Phosphorylation at this site increases the single channel conductance of AMPA receptors expressed in human embryonic kidney (HEK) 293 cells by about 40% (Roche et al. 1996; Derkach et al. 1999) and promotes the insertion of AMPA receptors into the postsynaptic membrane of glutamatergic synapses in cultured rat hippocampal neurons (Esteban et al. 2003). The latter process probably constitutes the final step in the generation of activity-dependent increases in synaptic strength at glutamatergic synapses and is essential for the expression of hippocampal LTP (Zamanillo et al. 1999; Mack et al. 2001). Incorporation of GluR-A into AMPA receptor channels hence enables AMPA receptors to function as endpoints in the generation of activity-dependent changes in synaptic transmission (Fig. 2).

The prominent expression of this "plasticity-permitting" GluR-A fits nicely with the fact that among the various afferent inputs to the spinal cord, input from nociceptors is the one that is most sensitive to plastic changes and central sensitization (Wall and Woolf 1984; Cook et al. 1987). Similar to tetanic stimulation in the hippocampus, intense nociceptive input to the dorsal horn leads to phosphorylation of GluR-A at Ser831 (and Ser845; Fang et al. 2003; Nagy et al. 2004). Furthermore, recruitment of GluR-A to the neuronal plasma membrane via an activity- and CaMKII-dependent process has recently been demonstrated in

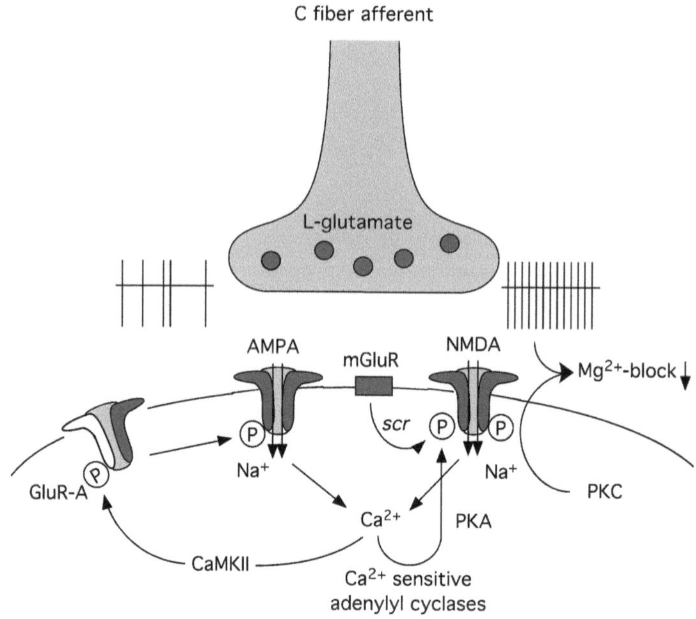

Fig. 2 Activity-dependent changes at glutamatergic synapses in the superficial dorsal horn. Low-frequency activity of C fibers mainly activates postsynaptic (±)-α-amino-3-hydroxy-5-methylisoxazole-4-propionic acid (*AMPA*) receptor channels. N-methyl-D-aspartate (*NMDA*) receptors remain inactive due to their blockade by extracellular Mg^{2+}. Under conditions of exaggerated C-fiber activity, glutamate released from C-fiber terminals also activates NMDA receptors and triggers Ca^{2+} influx through NMDA and, if present, through Ca^{2+}-permeable AMPA receptor channels into the postsynaptic neuron. The subsequent increase in intracellular free Ca^{2+} activates CaMKII, which phosphorylates GluR-A, leading to its translocation to the subsynaptic membrane and to an increase in channel conductance. Other kinases, including protein kinase A (Zou et al. 2002), protein kinase C (Chen and Huang 1992), and *scr* (Guo et al. 2004), activated by metabotropic glutamate receptors also phosphorylate NMDA receptor subunits and probably facilitate their activation. Phosphorylation by protein kinase C reduces the Mg^{2+} block and thereby further facilitates NMDA receptor activation (Chen and Huang 1992).

response to intracolonic instillation of capsaicin (Galan et al. 2004). It is therefore probably not too farfetched to speculate that phosphorylation of GluR-A also contributes to activity-dependent pain sensitization, or wind-up (Mendell 1966), seen during intense and prolonged nociceptive input to the dorsal horn. In light of these findings, it is not surprising that a recent study indeed demonstrated that mice lacking the GluR-A subunit exhibit reduced nociceptive sensitization in tests of tonic nociceptive stimulation (Hartmann et al. 2004). These results suggest that the generation of central nociceptive sensitization and hippocampal LTP have at least some basic events in common (for a review, see Ji et al. 2003).

Ca^{2+}-permeable AMPA receptors

Despite the similarities discussed above, AMPA receptors in the dorsal horn exhibit a number of rather peculiar features that may also be important for central nociceptive sensitization. Perhaps most striking among these is an unusually high number of Ca^{2+}-permeable AMPA receptors (Engelman et al. 1999), which reside both on γ-aminobutyric acid (GABA)ergic interneurons and on excitatory projection neurons (Albuquerque et al. 1999). Because Ca^{2+} plays an important role in synaptic plasticity, these receptors may critically contribute to LTP-like phenomena in the spinal cord (Gu et al. 1996). The low Ca^{2+} permeability of most AMPA receptors results from the insertion of the edited form of the GluR-B subunit in the channel complex. In most GluR-B transcripts, a certain adenosine residue is deaminated, leading to an exchange of glutamine to arginine at a critical position in the pore-forming M2 segment (Seeburg et al. 1998). It is not clear whether the high number of Ca^{2+}-permeable AMPA receptors in the dorsal horn results from a low expression of GluR-B (in relation to, for example, GluR-A) or from incomplete RNA editing. Despite this uncertainty, Ca^{2+}-permeable AMPA receptors seem to play an important role in central nociceptive sensitization. Both pharmacological and genetic evidence suggest that Ca^{2+} influx through dorsal horn AMPA receptors can trigger plastic changes in dorsal horn nociceptive circuits. Intrathecal injection of Joro spider toxin, which selectively blocks Ca^{2+}-permeable AMPA receptors, prevents the generation of mechanical allodynia in response to burn injury in rats (Sorkin et al. 1999, but see also Stanfa et al. 2000), and mice lacking the GluR-B subunit show not only increased Ca^{2+} influx in the superficial dorsal horn assessed by cobalt uptake but also a facilitation of nociceptive responses in the formalin test (Hartmann et al. 2004).

Presynaptic glutamate receptors

Ionotropic glutamate receptors are widely considered as sole postsynaptic sensors for glutamate, the primary function of which is the transmission of electric signals between neurons across synapses. However, all three types of ionotropic glutamate receptors (AMPA, kainate, and NMDA receptors) are also expressed in presynaptic axon terminals. This presynaptic expression is perhaps nowhere more prominent than in the spinal cord dorsal horn (Liu et al. 1994; Tachibana et al. 1994; Popratiloff et al. 1996; Hwang et al. 2001; Lu et al. 2002). The functional significance of these presynaptic receptors has been investigated for kainate and NMDA receptors. Activation of presynaptic kainate receptors located on primary afferent terminals inhibited glutamate release (Kerchner et al. 2001b), whereas those located at presynaptic terminals of inhibitory interneurons facilitated action-potential-independent release of GABA and glycine (Kerchner et al. 2001a). Similar results have been reported for presynaptic NMDA receptors. Liu et al. (1997) found that intrathecally injected NMDA triggered the release of substance P and glutamate from primary afferents, a process that may promote sensitization. However, more recently, Bardoni et al. (2004) provided evidence that activation of presynaptic NMDA receptors can also inhibit glutamate release. These apparently conflicting results are not as surprising as they seem at first glance. Whether the activation of a depolarizing presynaptic receptor may facilitate or inhibit transmitter release depends on the magnitude and steepness of depolarization. A similar paradox is discussed below for GABA-mediated primary afferent depolarization, which can either reduce transmitter release—as long as it remains subthreshold—or give rise to pronociceptive dorsal root reflexes when it becomes suprathreshold (Willis 1999).

NMDA receptors

The pivotal role of NMDA receptors in the induction of synaptic plasticity in many CNS areas has been so widely accepted that this review will focus on some aspects that might be particularly relevant to nociceptive processing in the dorsal horn. NMDA receptor activation contributes to pain sensation in at least two respects: first, it is important for the transmission of acute pain through the spinal cord, as drugs that block NMDA receptors, such as ketamine, can produce profound analgesia, and, second, it is required for central sensitization and LTP-like phenomena in the dorsal horn (Liu and Sandkühler 1995). Plastic changes in excitatory synaptic transmission onto dorsal horn projection neurons in response to intense C-fiber input probably contribute to the generation of enduring hyperalgesia after tissue trauma (Sandkühler 2000). Although the high firing frequency used in initial in vitro studies in spinal cord slices is not observed in C fibers in vivo, other more physiological paradigms used more recently have yielded similar results (Sandkühler and Liu 1998). Interestingly, the susceptibility to long-lasting changes in synaptic strength varies considerably within dorsal horn neurons. It is most prominent in spinothalamic projection neurons expressing NK1 (substance P) receptors (Ikeda et al. 2003). Because a pivotal role of NK1 receptor-positive neurons for a number of different forms of pain has been demonstrated previously (Mantyh et al. 1997; Nichols et al. 1999; Khasabov et al. 2002), this result endorses an important contribution of dorsal horn synaptic plasticity to the development of chronic pain syndromes.

One important difference between central sensitization in pain pathways and "typical" LTP is that the latter is input-specific. In the hippocampus, increases in synaptic strength are largely restricted to those synapses that have been active during the conditioning stimulation. This feature corresponds to *specificity,* one of three characteristics that have been postulated by the Canadian psychologist Donald Hebb for neuronal correlates of associative learning (Hebb 1966). Unlike hippocampal LTP, central sensitization in pain pathways is not fully specific. Intense stimulation of cutaneous C fibers elicits secondary (central) hyperalgesia in an area significantly exceeding the field of conditioning stimulation, indicating that primary afferent fibers not activated during the conditioning stimulus can become sensitized. Yet intense C-fiber input can even potentiate $A\beta$ fiber (touch)-evoked responses of dorsal horn neurons (e.g., Coderre and Melzack 1987). Hence, central sensitization is also heterosynaptic and could thus be either totally unrelated to LTP or could be made less specific by mechanisms involving LTP in the dorsal horn.

One such possibility could be the spread of diffusible messengers released during intense nociceptive stimulation. Glycine is one such diffusible messenger required for NMDA receptors to become fully active (Johnson and Ascher 1987; Kleckner and Dingledine 1988). In the first years after the discovery of the glycine binding site at NMDA receptors, it was unclear whether glycine binding could contribute to NMDA receptor modulation in vivo or whether this site was permanently saturated by micromolar concentrations of glycine in the cerebrospinal fluid (Westergren et al. 1994). During the last few years, however, increasing evidence has accumulated indicating that plasma membrane glycine transporters, in particular the glial glycine transporter GlyT1, can lower extracellular glycine concentrations in the vicinity of NMDA receptors to subsaturating concentrations (Bergeron et al. 1998; Berger et al. 1998; Gabernet et al. 2005). Perhaps nowhere else in the CNS is a contribution of synaptically released glycine to NMDA receptor facilitation more likely to occur than in the spinal cord, where glycinergic neurons are very abundant. Using nocistatin (Okuda-Ashitaka et al. 1998), a peptide that in the dorsal horn selectively inhibits the release of glycine (and GABA; Zeilhofer et al. 2000), Ahmadi et al. (2003) demonstrated that during intense nociceptive stimulation, glycine synaptically released in the dorsal horn can overcome reuptake by glycine transporters and reach neighboring NMDA receptors to facilitate their activation through a process called spillover. Increased activation of NMDA receptors through glycine spillover apparently contributed to sustained formalin-induced pain behavior and to neu-

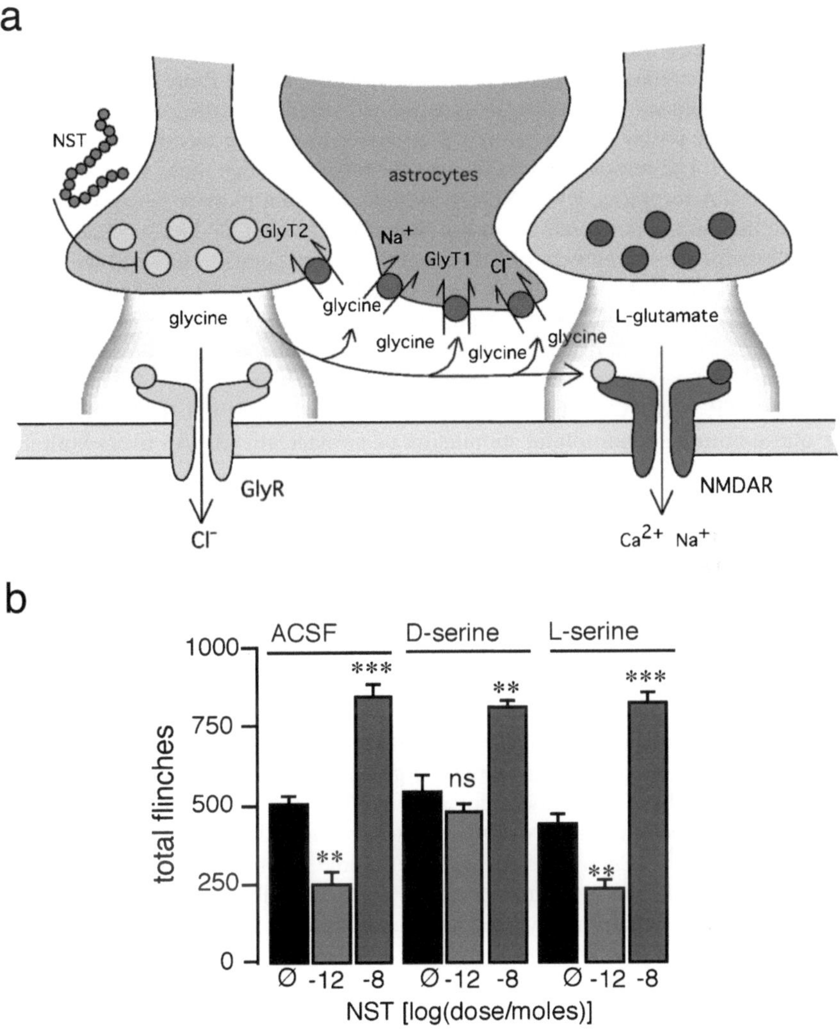

Fig. 3 Spillover of synaptically released glycine facilitates NMDA receptor activation in the dorsal horn. **a** Under conditions of exaggerated nociceptive input to the dorsal horn, glycine released from inhibitory interneurons or descending glycinergic fiber tracts can escape the synaptic cleft of the glycinergic synapses and reach nearby NMDA receptors to facilitate their activation (Ahmadi et al. 2003). **b** Evidence for a physiological role of this process in central sensitization has been obtained in the rat formalin test. Nocistatin (*NST*), a peptide that in the dorsal horn selectively reduces the release of glycine (and GABA), evokes pro- or antinociceptive effects after intrathecal injection, depending on the dose injected. The antinociceptive effect is specifically antagonized by D-serine, an activator of the glycine binding site of NMDA receptors, indicating that nocistatin suppressed nociception by reducing the availability of glycine at NMDA receptors. Artificial cerebrospinal fluid (*ACSF*) and L-serine were ineffective (for details, see Ahmadi et al. 2003). $**P \leq 0.01$; $***P \leq 0.001$

ropathic pain in the chronic constriction injury in rats (Muth-Selbach et al. 2004). During intense nociceptive stimulation, spillover of glycine might thus promote the potentiation of synaptic input at sites not fully activated during the conditional stimulation (Fig. 3).

An interesting hypothesis that has again originated from previous findings in the hippocampus postulates that pathological pain might come from the functional activation of previously silent excitatory synapses in the dorsal horn (Li and Zhuo 1998). Early in hippocampal development, glutamatergic synapses are silent at resting potential but can be activated when the postsynaptic neuron is depolarized to positive membrane potentials (Durand et al. 1996). The reason behind this unusual behavior is that these silent synapses lack functional AMPA receptors, while NMDA receptors present in these synapses are blocked at negative membrane potentials by extracellular Mg^{2+}. An LTP-like mechanism appears to be required for the recruitment of functional AMPA receptors to these synapses. Similar silent synapses have also been found in the neonatal dorsal horn (Li and Zhuo 1998) but not in adult animals (Baba et al. 2000a). A possible contribution of the activation of previously silent synapses to the generation of pathological pain in the adult therefore remains to be demonstrated.

Besides LTP, NMDA receptor-dependent long-term depression (LTD) can be evoked in the dorsal horn by conditioning stimulation of primary afferent Aδ fibers both in slices (Randic et al. 1993; Sandkühler et al. 1997) and in vivo (Liu et al. 1998). Long-lasting depression of Aδ fiber-mediated postsynaptic responses can also be elicited by activation of group I metabotropic glutamate receptors and subsequent phospholipase C activation (Chen et al. 2000). Very recently, Klein et al. (2004) demonstrated perceptual correlates for LTP and LTD in human volunteers after cutaneous electrical stimulation.

It should finally be noted that NMDA receptors are not only important for the induction of LTP and LTD by permitting the necessary postsynaptic Ca^{2+} increase but are themselves regulated by neuronal activity and protein kinases. Peripheral inflammation and tonic nociceptive stimulation induce PKA-dependent phosphorylation of the NR1 (Zou et al. 2002) and *scr*-dependent phosphorylation of the NR2B subunit (Guo et al. 2004) of NMDA receptors. *scr* is activated by group I metabotropic glutamate receptors, a process that explains the contribution of these receptors to dorsal horn LTP (Azkue et al. 2003). Protein kinase C-dependent phosphorylation of NMDA receptors reduces their susceptibility to blockade by extracellular Mg^{2+} (Chen and Huang 1992). Some mechanisms of activity-dependent plasticity in the dorsal horn are summarized in Fig. 2. A possible contribution of these phosphorylation events to inflammatory pain is discussed below.

AMPA and NMDA receptors as targets for analgesic drugs

Although both competitive [e.g., 6-cyano-7-nitroquinoxaline-2,3-dione (CNQX)] and non-competitive (e.g., GYKI 52466) AMPA receptor antagonists exhibit antinociceptive properties in a variety of animal models of pain (Szekely et al. 2002), their use as analgesics is severely hampered by their widespread action in the CNS. However, in contrast to most CNS areas, where fast excitatory neurotransmission is almost exclusively mediated by AMPA receptors, kainate receptors composed of GluR-5, GluR-6, and GluR-7 significantly contribute to primary afferent nociceptive transmission in the spinal cord (Li et al. 1999). Kainate receptor antagonists exert antinociceptive properties in different neuropathic pain models and in the formalin test (for a review, see Ruscheweyh and Sandkühler 2002). Whether selective kainate receptor blockers are better tolerated than unspecific ones is unknown at present.

N-methyl-D-aspartate receptor antagonists have attracted significantly more attention as possible analgesics than AMPA receptor blockers have. This is probably for two reasons. First, NMDA receptors are typically not required for fast excitatory synaptic transmission under basal conditions. Second, the pivotal role of NMDA receptors for synaptic plasticity has raised the hope that the "preemptive" blockade of NMDA receptors might prevent the generation of chronic pain after tissue injury. Blockers of NMDA receptors are clearly analgesic, as exemplified by the intravenous anesthetic ketamine and a variety of experimental drugs (e.g., Qian et al. 1996). However, sedation and the impairment of motor coordination preclude their long-term use as analgesics. Competitive antagonists at the glycine-binding site (e.g., L-701324) of NMDA receptors show less maximal effect compared with channel blockers (e.g., ketamine or MK-801) or competitive antagonists at the glutamate binding site (e.g., AP-5), but they are probably also less efficient as analgesics. Finally, specific antagonists of NMDA receptors containing the NR2B subunit (such as ifenprodil), which is preferentially expressed in the dorsal horn, have raised new hope but still have not begun clinical application (Chizh et al. 2001).

Disinhibition of spinal nociception: a common mechanism in inflammatory and neuropathic pain

As outlined above, AMPA and NMDA receptors are critically involved in activity-dependent changes in spinal nociceptive processing. Their contribution to the pathogenesis of inflammatory pain or to pain originating from peripheral nerve damage, however, is less clear. Accordingly, recent evidence points to a critical role of disinhibition—that is, a reduction in the activity of glycinergic and GABAergic neurons and receptors—rather than direct exaggerated excitation as the dominant source of inflammatory and neuropathic pain.

Glycinergic and GABAergic inhibition in the dorsal horn

Inhibitory synaptic transmission onto superficial dorsal horn neurons probably originates from different sources. Local inhibitory interneurons can be activated by primary nociceptive afferents (Narikawa et al. 2000) or by descending antinociceptive fiber tracts. Inhibitory input can also directly come from descending GABAergic and glycinergic fiber tracts projecting from the rostral ventromedial medulla to the dorsal horn (Antal et al. 1996). GABA and glycine open ligand gated ion channels, which permit the permeation of chloride and, to a lesser extent, bicarbonate ions through the plasma membrane. In most neurons, both transmitters inhibit neuronal activation by hyperpolarizing the cell membrane and by activating a shunting conductance, which impairs the propagation of excitatory postsynaptic potentials along the dendrite of neurons. Early in postnatal development, both transmitters are coreleased from the same vesicles. In the adult a most likely postsynaptic specialization occurs, which makes mixed GABA/glycinergic postsynaptic events less frequent (Keller et al. 2001). Several lines of evidence suggest that synaptic inhibition in the superficial dorsal horn is mainly mediated by glycine, whereas synaptically released GABA primarily acts on presynaptic $GABA_B$ (Chery and de Koninck 2000) and extrasynaptic $GABA_A$ receptors to provide tonic inhibition (Chery and de Koninck 1999). It has repeatedly been speculated that the proper functioning of this inhibitory input is essential to prevent the generation of painful sensations by normally innocuous stimuli. It has long been known that pharmacological removal of inhibitory GABAergic or glycinergic inhibition contributes to central sensitization in the spinal cord (Sivilotti and Woolf 1994). More recent publications now indicate that a reduction in the inhibitory tone in the spinal cord dorsal horn by endogenous mediators underlies several forms of pathological pain.

Actions of prostaglandins on synaptic transmission in the dorsal horn

Tissue damage and inflammation trigger the release of arachidonic acid from phospholipids of the cell membrane through the activation of phospholipase A2. Arachidonic acid is then converted by constitutively expressed cyclooxygenase-1 (COX-1) and inducible cyclooxygenase-2 (COX-2) into the two prostaglandin (PG) precursors PGG2 and PGH2. Tissue-specific isomerases or prostaglandin synthases further process arachidonic acid into the biologically active prostaglandins (PGE2, PGD2, PGI2, and PGF2α) and into thromboxane A2. For a long time it has generally been assumed that prostaglandins sensitize the nociceptive system only at the level of the peripheral nociceptor. However, during the last 15 years increasing evidence has accumulated indicating that prostaglandins can also cause hyperalgesia in the CNS, in particular in the spinal cord dorsal horn. Peripheral inflammation induces the expression of COX-2 and of the microsomal prostaglandin E synthase (mPGES) in the spinal cord and possibly also elsewhere in the CNS (Beiche et al. 1996; Samad et al. 2001; Guay et al. 2004; Kamei et al. 2004). Inhibition of prostaglandin formation in the spinal cord by cyclooxygenase inhibitors or nonsteroidal antiinflammatory drugs (NSAIDs) produces antinociception in a variety of pain models, while injection of prostaglandin E2 into the spinal canals of mice and rats causes profound hyperalgesia to thermal and mechanical stimuli and allodynia (for a review, see Vanegas and Schaible 2001). Despite this overwhelming evidence for a central pronociceptive action of prostaglandins, the molecular mechanisms of central inflammatory hyperalgesia have long remained elusive. A better understanding of which prostaglandins and which prostaglandin receptors are responsible for central pain sensitization is, however, essential for the development of novel better-tolerated analgesics.

Prostaglandin E2 exerts its cellular effects through the activation of four different types of G-protein-coupled rhodopsin-like receptors, called EP1 through EP4, which differ in their tissue distribution and signal transduction (Narumiya et al. 1999). Electrophysiological studies in both the intact spinal cords of rats and in isolated slice preparations have led researchers to propose several different possible mechanisms of action (Fig. 4). These include a prostaglandin-mediated increase in the release of the excitatory transmitter glutamate (Minami et al. 1999), an increased responsiveness of postsynaptic AMPA or NMDA receptors (Vasquez et al. 2001; Bär et al. 2004), a direct depolarization of deep dorsal horn neurons (Baba et al. 2001), and an inhibition of postsynaptic inhibitory (strychnine-sensitive) glycine receptors (Ahmadi et al. 2002). Among those, the direct depolarization of deep dorsal horn neurons and the inhibition of glycine receptors have been studied in detail. Baba et al. (2001) demonstrated that low micromolar concentrations of prostaglandin E2 depolarize a subpopulation of neurons mainly but not exclusively located in the deep dorsal horn through the activation of a cationic conductance by EP2 or EP2-like receptors. Ahmadi et al. (2002) found that low nanomolar concentrations of prostaglandin E2 inhibited glycinergic neurotransmission through a postsynaptic mechanism involving EP2 receptors and the activation of protein kinase A. Interestingly, both groups found no evidence for a direct effect of prostaglandin E2 on either glutamate release or on the responsiveness of postsynaptic glutamate receptors. The identification of the glycine receptor subunit inhibited by prostaglandin E2/protein kinase A and the advent of genetically engineered mice deficient in this glycine receptor subunit has recently allowed the relevance of both mechanisms for inflammatory hyperalgesia in vivo to be determined.

Native glycine receptors in the adult are heteropentameric protein complexes (for recent reviews, see Lynch 2004 and Legendre 2001). Five different glycine receptor subunits, termed GlyRα1 through GlyRα4 and GlyRβ, are known. GlyRα subunits bind glycine and

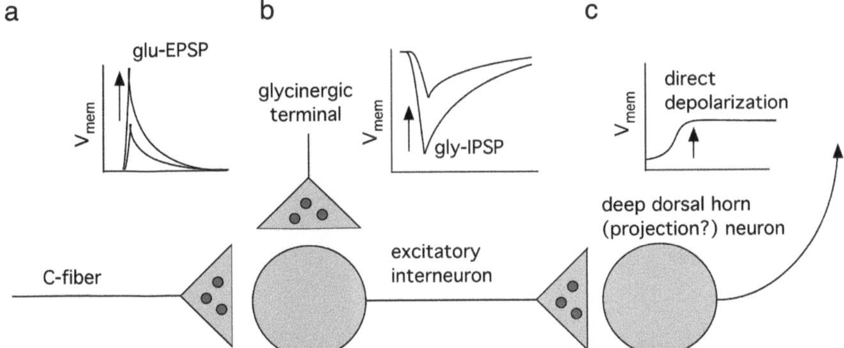

Fig. 4 Synaptic mechanisms proposed for the pronociceptive defects of prostaglandin E2. **a** Prostaglandin E2 facilitates the release of L-glutamate from nociceptive C fibers (Minami et al. 1999). **b** Prostaglandin E2 blocks inhibitory glycine receptors in the superficial dorsal horn (Ahmadi et al. 2002, Harvey et al. 2004). **c** Prostaglandin E2 directly depolarizes deep dorsal horn neurons (Baba et al. 2001). Experimental evidence for a contribution to inflammatory pain in vivo has so far been obtained only for the inhibition of glycine receptors (Harvey et al. 2004, Reinold et al. 2005).

are capable of forming functional homomeric channels, while the so-called structural GlyRβ subunit confers subsynaptic clustering through an interaction with the postsynaptic protein gephyrin. The most prevalent isoform of glycine receptors in the adult is composed of GlyRα1 and GlyRβ subunits. GlyRα3 is another much less prevalent adult glycine receptor isoform; GlyRα2 is widely believed to be an embryonic and juvenile isoform in most parts of the CNS; and GlyRα4 may even be a pseudogene in humans.

Harvey et al. (2004) reconstituted the inhibitory effect of prostaglandin E2 on glycinergic membrane currents in a heterologous expression system. After cotransfection of HEK293 cells with EP2 receptors and different glycine receptor subunits, it became apparent that currents through glycine receptors containing the GlyRα3 subunit were inhibited by prostaglandin E2, whereas GlyRα1 was not. As expected from the experiments in spinal cord slices, this inhibition was prevented by perfusion of the recorded neurons with a PKA inhibitor peptide. Further experiments revealed that PKA inhibited GlyRα3-containing glycine receptors most likely through the phosphorylation of a serine residue (Ser346) located in the long intracellular loop between transmembrane segments S3 and S4. Figure 5a shows a schematic representation of the synaptic signal transduction. Interestingly, glycine receptor subunits show a distinct pattern of expression in the spinal cord. GlyRα1 and GlyRβ are found throughout the grey matter spinal cord, while the GlyRα3 subunit is expressed only in the superficial layers of the dorsal horn where most nociceptive afferents terminate (Fig. 5b; Harvey et al. 2004) and where prostaglandin E2-mediated inhibition of glycinergic neurotransmission had been observed in slices (Ahmadi et al. 2002).

The availability of mice lacking GlyRα3 permitted the relevance of this pathway for inflammatory pain sensitization in vivo to be determined. GlyRα3-deficient mice not only lack inhibition of glycinergic neurotransmission by prostaglandin E2 but also show a dramatic reduction in the pronociceptive effects of spinal prostaglandin E2 in vivo. In addition, these mice recover much faster from inflammatory hyperalgesia following subcutaneous injection of the yeast extract zymosan A or of complete Freund's adjuvant than their wild-type littermates do. Mice lacking the EP2 subtype of prostaglandin E2 receptors exhibit a nearly identical phenotype after subcutaneous zymosan A injection (Fig. 5c,d) and also lack prostaglandin E2-induced inhibition of glycinergic neurotransmission (Reinold et al. 2005;

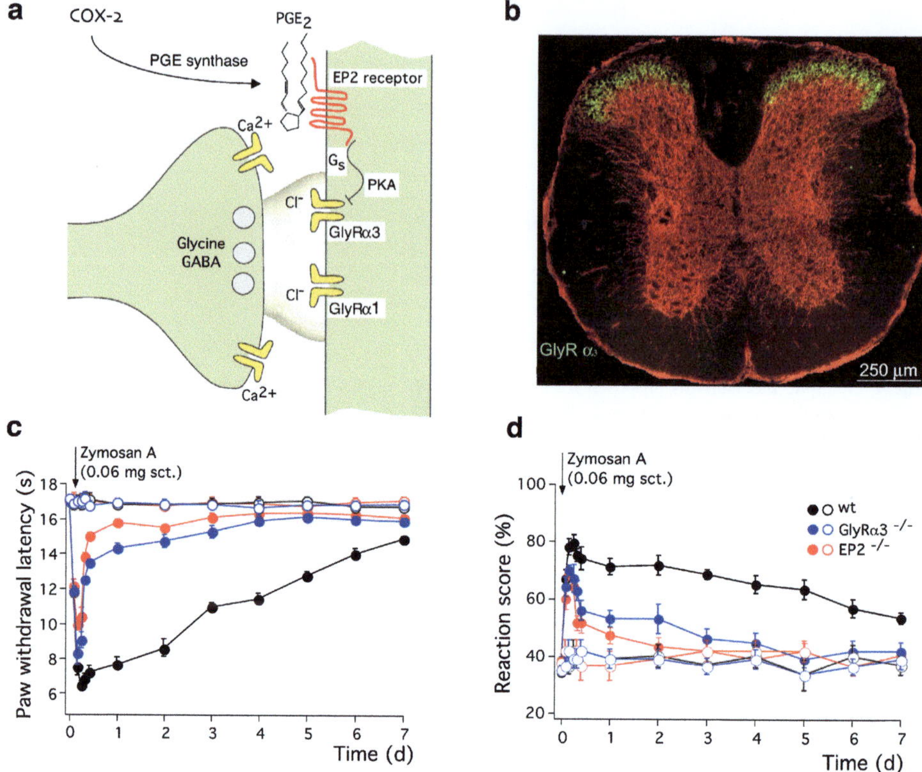

Fig. 5 Synaptic disinhibition underlies inflammatory hyperalgesia in the dorsal horn. Cyclooxygenase-2 (*COX-2*) is induced in the spinal cord in response to peripheral inflammation. **a** Prostaglandin E2 (*PGE₂*) acts on postsynaptic EP2 receptors and leads to protein kinase A (*PKA*)-dependent phosphorylation and inhibition of the glycine receptor subunit α3 (*GlyRα3*). **b** GlyRα3 is distinctly expressed in the superficial dorsal horn where most nociceptive afferents terminate. **c, d** Mice lacking either the EP2 receptor or the GlyRα3 subunit exhibit dramatically reduced inflammatory hyperalgesia to **c** thermal and to **d** mechanical stimuli. *Filled symbols* indicate zymosan A-injected paw, and *open circles* indicate contralateral noninjected paw. Thermal hyperalgesia was assessed as latency of paw withdrawal in response to exposure to a defined radiant heat stimulus. Mechanical sensitization was assessed in response to stimulation with calibrated von Frey filaments (for details, see Depner et al. 2003).Data inpart taken from Reinold et al. 2005

Zeilhofer 2005). These findings correspond nicely to previous observations by Malmberg et al. (1997), who have reported that mice lacking the neuronal isoform of protein kinase A show reduced nociceptive sensitization after intrathecal injection of prostaglandin E2. Protein kinase A-dependent phosphorylation and inhibition of GlyRα3 in response to EP2 receptor activation appears, therefore, as the dominant mechanism of central inflammatory pain sensitization (Fig. 5a). This disinhibition renders excitatory input more effective and thereby probably facilitates the induction of activity-dependent plasticity through NMDA receptor activation, eliciting many of the molecular events described in the previous section. In light of these findings, it is reasonable to propose that the antihyperalgesic action of cyclooxygenase inhibitors is mainly due to the inhibition of this process. From a therapeutic perspective, future EP2 receptor antagonists and drugs enhancing the function of GlyRα3 might be considered as centrally acting nonopioidergic antihyperalgesic agents.

GABA$_A$ receptors on the central terminals of primary nociceptive afferent nerve fibers

Under physiological conditions, activation of GABA$_A$ receptors hyperpolarizes the postsynaptic neuron through an influx of chloride ions in almost all areas of the adult CNS. In this respect the central terminals of primary afferent nerve fibers are important exceptions. Due to an unusually high intracellular chloride concentration, the opening of GABA$_A$ receptor channels in these terminals induces a depolarization, which under "normal" conditions inhibits transmitter release. These presynaptic GABA$_A$ receptors residing on the spinal terminals of primary afferents are probably activated by local dorsal horn interneurons, which make axoaxonic synapses with other primary afferent terminals. It has been proposed that these inhibitory interneurons are excited by low threshold mechanoreceptors (Aβ fibers) and contact the central terminals of primary afferent nociceptors (for a review, see Willis 1999). Peripheral Aβ fiber activation would thereby *inhibit* nociceptive transmission. It has further been proposed that inflammation and perhaps neuropathy could increase primary afferent depolarization to become suprathreshold and to elicit action potentials. These action potentials would travel in anterograde and retrograde directions to *elicit* transmitter release both at the central and peripheral terminals of nociceptors. Under these conditions, Aβ fiber activation would give rise to so-called dorsal root reflexes (Rees et al. 1995). This or a similar mechanism could contribute to central hyperalgesia and allodynia (Cervero and Laird 1996). It may also contribute to heterosynaptic potentiation, a typical feature of central pain sensitization. If these GABAergic interneurons are also contacted by nociceptors, dorsal root reflexes could also explain the spread of neurogenic inflammation and hyperalgesia beyond the site of peripheral stimulation.

Synaptic disinhibition in the dorsal horn as a possible source of neuropathic pain

Neuropathic pain results from the damage of neurons or nerves through trauma, poisoning, or metabolic diseases. Damage of peripheral nerves through mechanical injury or metabolic diseases is probably the most frequent cause of neuropathic pain. As in inflammatory pain, the initial focus of research was on peripheral processes. It could be demonstrated that damaged nerves become spontaneously active (Sheen and Chung 1993) and that irregular electric contacts may be formed between nerve fibers (e.g., Seltzer and Devor 1979). Both processes probably explain some aspects of neuropathic pain, but they cannot fully account for all aspects of neuropathic pain syndromes. Central processes have therefore become the center of present research. During the last decade, a number of cytokines and other mediators have been identified that contribute to the development of pain after peripheral nerve injury (Sommer 2003). However, insight into how these mediators affect sensory information processing has only very recently been gained.

One of the most unpleasant sensations in neuropathic pain is the painful sensation of stimuli that are normally not sensed as painful, such as light touch. A consistent finding of many studies addressing this issue appears to be a loss of the inhibitory tone in the spinal cord dorsal horn. Although it might be speculated that the production of cytokines in the CNS after peripheral nerve trauma stimulates prostaglandin production, a significant contribution of cyclooxygenase-2 to neuropathic pain appears rather unlikely (e.g., Broom et al. 2004). Mice deficient in the EP2 receptor develop normal hyperalgesia and allodynia in the chronic constriction injury model of neuropathic pain, although they completely lack the pronociceptive effects of spinally administered prostaglandin E2 (Hösl et al., in preparation). However, several recent publications suggest that relief from GABAergic or glycinergic inhibition of spinal nociceptive neurons through prostaglandin-independent mechanisms contributes to neuropathic pain. Possible actions include an inhibition of glycine or GABA release from inhibitory interneurons; a reduction in the transmembrane chloride gradient, rendering inhibition by glycinergic and GABAergic synaptic input less efficient; and a loss of inhibitory innervation due to a selective death of GABAergic or glycinergic interneurons.

There is indeed experimental evidence for all three possibilities, but their contribution to pain sensitization is far from being fully clear. A reduction in the transmembrane chloride gradient in dorsal horn neurons following peripheral nerve injury has recently been reported by Coull et al. (2003). Peripheral nerve trauma induces a transsynaptic reduction in the expression of the potassium chloride exporter KCC2 in dorsal horn neurons and thereby shifts the chloride equilibrium potential to more depolarized values. This shift reduces the inhibitory effect of GABA and glycine receptor activation, yet it might even cause glycinergic or GABAergic input to become excitatory.

Another extensively discussed report suggests that peripheral nerve injury induces a specific loss of spinal inhibitory GABAergic neurotransmission in the dorsal horn of rats in the chronic constriction injury model and the spared nerve injury model of neuropathic pain (Moore et al. 2002a). This original report has suggested that the loss of GABAergic input was due to the selective apoptotic death of GABAergic interneurons. Subsequent studies have, however, shown that such a loss is at least unnecessary for the development of thermal hyperalgesia in the chronic nerve injury model of neuropathic pain (Polgar et al. 2003, 2004).

Inhibitory neurotransmitter receptors: targets for novel analgesic drugs?

Given that a relief from glycinergic and GABAergic inhibition underlies inflammatory and neuropathic pain, pharmacological enhancement of the function of these neurotransmitter receptors should be considered a future therapeutic option. Bearing in mind the pivotal role of the glycine receptor $\alpha 3$ subunit in the spinal processing of nociceptive signals, a selective agonist at GlyR$\alpha 3$ or, even better, a positive allosteric modulator would be desirable. Unfortunately, so far no specific agonists for GlyR$\alpha 3$ have been identified, and only a few agents are known to potentiate GlyR-mediated currents, with most of them lacking receptor specificity (Breitinger and Becker 1998; Laube et al. 2002).

Unlike glycine receptors, GABA$_A$ receptors are extensively used as therapeutic targets. The classical benzodiazepines, which enhance GABA responses at benzodiazepine-sensitive GABA$_A$ receptors, cause sedation, anxiolysis, and central muscle relaxation and are anticonvulsive. Antinociceptive actions of benzodiazepines have been repeatedly described in animal models of pain, in particular after spinal injection (e.g., Goodchild and Serrao 1987), and a few studies have reported analgesic effects of agonists at GABA$_A$ receptors in human patients (for reviews, see Jasmin et al. 2004 and Krogsgaard-Larsen et al. 2004). However, classical benzodiazepines are not routinely used as analgesics. Underlying reasons may include the accompanying sedation, which probably occurs at lower doses than that needed for analgesia, and possible pronociceptive effects of benzodiazepines at supraspinal sites (Tatsuo et al. 1999), which may counteract spinal antinociception. This may not necessarily be the case for future subunit-specific benzodiazepines. The generation of "knock-in" mice carrying point mutations at the benzodiazepine binding sites of the different GABA$_A$ receptor subunits has proven that the different actions of benzodiazepines can be assigned to different $GABA_A$ receptor subunits (Rudolph and Möhler 2004). This has already led to the discovery of nonsedative anxiolytic "benzodiazepines" (Löw et al. 2000; McKernan et al. 2000). Future research will have to determine whether subunit-specific drugs can reverse the loss of neuronal inhibition occurring during inflammation and neuropathy.

Modulators of synaptic transmission not discussed above

In the previous sections, I have concentrated on synaptic mechanisms underlying the generation of pathological pain through activity-dependent sensitization, inflammation, and neuropathy. It should, however, be noted that the conscious perception of pain is strongly dependent on our psychological context. The biological correlate for this probably lies in the large number of neuromodulators that are released under various conditions in the CNS and that interfere with nociceptive processing at different levels of integration. In the following section I will briefly summarize synaptic effects in the dorsal horn of mediators and receptors not discussed in the previous sections, which modulate neurotransmission in the dorsal horn either through presynaptic (Table 1) or postsynaptic mechanisms (Table 2).

Table 1 Presynaptic modulators in the dorsal horn

Target of modulation	Modulator (receptor)	Receptor	Effect	Reference
Glutamate	Opioids	μ, δ	Inhibition	Light and Willcockson 1999; Kohno et al. 1999
	N/OFQ	NOP	Inhibition	Liebel et al. 1997; Zeilhofer et al. 2000; C. Luo et al. 2002a
	Norepinephrine	$\alpha2A$	Inhibition	Pan et al. 2002; Kawasaki et al. 2003
	GABA	$GABA_B$	Inhibition	Ataka et al. 2000
	Serotonin	-HT-1D	Inhibition	Travagli and Williams 1996; Garraway and Hochman 2001
		-HT3	Facilitation/inhibition	Khasabov et al. 1999; Hori et al. 1996
	Acetylcholine	Nicotine	Inhibition	Takeda et al. 2003; Kiyosawa et al. 2001
	Cannabinoids	CB-1	Inhibition	Harris et al. 2000; Kelly and Chapman 2001; Morisset and Urban 2001
	ATP	P2X	Facilitation	Gu and MacDermott 1997; Li et al. 1998
	Adenosine, ADP	P2Y	Inhibition	Lao et al. 2004
	Prostaglandin E2	??	Facilitation	Minami et al. 1999
GABA/glycine	Opioids	μ	Inhibition	Grudt and Henderson 1998
	Nocistatin	??	Inhibition	Zeilhofer et al. 2000; Ahmadi et al. 2001
	Norepinephrine	$\alpha2$	Inhibition	Baba et al. 2000b, 2000c
	Adenosine	A1	Inhibition	Yang et al. 2004; Hugel and Schlichter 2003
	GABA	$GABA_B$	Inhibition	Iyadomi et al. 2000
	Serotonin	P2X	Facilitation	Rhee et al. 2000; Jang et al. 2001; Hugel and Schlichter 2000

Table 2 Postsynaptic modulators in the dorsal horn

Target of modulation	Modulator	Effect	Mechanism	Site	Reference
AMPA	Neuronal activity, Ca^{2+}	Potentiation, incorporation into postsynaptic membrane	CaMKII-dependent phosphorylation of GluR-A	Probably ubiquitous superficial dorsal horn	Barria et al. 1997; Roche et al. 1996; Derkach et al. 1999; Fang et al. 2003; Nagy et al. 2004
NMDA	Opioids (?)	Potentiation	PKC-dependent reduction in Mg^{2+} block	Trigeminal nucleus	Chen and Huang 1992
	Glutamate (group I mGluRs) glycine	Potentiation	scr-dependent phosphorylation of NR2B	Dorsal horn	Guo et al. 2004
	Glycine	Potentiation	Direct binding to NR1	Probably ubiquitous	Johnson and Ascher 1987
				Superficial dorsal horn	Ahmadi et al. 2003
$GABA_A$	Serotonin	Potentiation	PKC	Lamina X	Xu et al. 1998
				Superficial dorsal horn	Li et al. 2000
Glycine	Norepinephrine (α2)	Potentiation	cAMP $\downarrow$, inhibition of PKA	Lamina X	Nabekura et al. 1999
	Serotonin	Potentiation	PKC	Lamina X	Xu et al. 1996
	Prostaglandin E2 (EP2)	Inhibition	cAMP $\uparrow$, PKA-dependent inhibition of GlyRα3	Superficial dorsal horn	Ahmadi et al. 2002; Harvey et al. 2004; Reinold et al. 2005

Opioid receptors

Opioids are the prototypes of centrally acting analgesics. Part of their analgesic activity results from inhibition of nociceptive transmission in the dorsal horn. At this site, activation of μ and δ opioid receptors hyperpolarizes nociceptive neurons, most likely through an activation of G-protein-coupled K^+ channels (Schneider et al. 1998; Mitrovic et al. 2003), and inhibits the release of glutamate from primary nociceptive afferents (Light and Willcockson 1999; Kohno et al. 1999), possibly through inhibition of voltage-gated Ca^{2+} channels (Schroeder and McClesky 1993; Taddese et al. 1995). The effects of Tyr-D-Ala-Gly-N-methyl-Phe-Gly-ol (DAMGO), a selective μ-opioid receptor agonist, on GABAergic or glycinergic inhibitory postsynaptic currents (IPSCs) in substantia gelatinosa neurons are controversial. Although Kohno et al. (1999) found no effect at all, Grudt and Henderson (1998) reported an inhibition of glycinergic and GABAergic postsynaptic currents via a presynaptic mechanism. Work with mice deficient in the individual subtypes of opioid receptors has shown that both desired analgesic and unwanted effects of morphine derive from μ-opioid receptor activation (Matthes et al. 1996), disappointing hopes that a separation of desired and unwanted opioid actions could be obtained through selective activation of a single opioid receptor subtype.

In 1994 a receptor closely related to the classical opioid receptors was identified and termed ORL-1 receptor (Mollereau et al. 1994). Activation of this receptor by its endogenous ligand called nociceptin (Meunier et al. 1995) or orphanin FQ (N/OFQ; Reinscheid et al. 1995) selectively inhibits glutamate release in the dorsal horn (Liebel et al. 1997; Zeilhofer et al. 2000; C. Luo et al. 2002a). Interestingly, N/OFQ has antiopioidergic activity at supraspinal sites, namely the nucleus raphe magnus (Grisel et al. 1996). At this site, classical opioids preferentially inhibit the release of GABA and thereby disinhibit descending antinociceptive fiber tracts (Basbaum and Fields 1984), whereas N/OFQ inhibits nearly all cell types (Vaughan et al. 2001). In behavioral experiments, N/OFQ exerts profound antinociception after spinal injection (e.g., Erb et al. 1997) but exerts antiopioidergic and hyperalgesic effects after injection at supraspinal sites (Meunier et al. 1995; Reinscheid et al. 1995). Mice lacking either the N/OFQ precursor peptide or the N/OFQ receptor exhibit increased nociceptive responses in tests of tonic chemical or inflammatory pain (Depner et al. 2003), suggesting that N/OFQ contributes to endogenous pain control in vivo. For a review focusing on the pain modulating action of N/OFQ, see Zeilhofer and Calò (2003).

Another peptide released from the N/OFQ precursor peptide is nocistatin (Okuda-Ashitaka et al. 1998). This peptide specifically inhibits GABA and glycine release in the dorsal part of the spinal cord, but it has no effect on glutamate release (Zeilhofer et al. 2000). After intrathecal injection, it increases nociceptive responses in the rat formalin test (Zeilhofer et al. 2000) and in the chronic constriction injury model of neuropathic pain (Muth-Selbach et al. 2004). Although nocistatin has been found in the cerebrospinal fluid of humans and animals (Lee et al. 1999), a contribution of endogenous nocistatin to thermal hyperalgesia evoked by inflammation (zymosan A) or to nociceptive responses in the formalin test could not be detected (Depner et al. 2003).

Norepinephrine and serotonin

Norepinephrine and serotonin (5-hydroxytryptamine, 5-HT) are important transmitters in antinociceptive fiber tracts descending from the raphe magnus nucleus to the dorsal horn and interfere with synaptic transmission in the dorsal horn in several ways. Norepinephrine reduces the release of glutamate from primary afferent terminals in the substantia gelatinosa (Pan et al. 2002; Kawasaki et al. 2003) and at the same time facilitates glycine release in the spinal cord dorsal horn (Baba et al. 2000b, 2000c). Both effects are mediated through the activation of $\alpha 2$ adrenoceptors and impair the transmission of nociceptive signals through the spinal cord. They probably underlie the analgesic effects of the $\alpha 2$ adrenoceptor agonist clonidine and related substances. Unfortunately, at least the inhibition of glutamate release is mediated by the $\alpha 2A$ subtype (Hunter et al. 1997; Lakhlani et al. 1997), which also causes sedation, disabling the pharmacological dissociation of sedative and analgesic effects of $\alpha 2$ adrenoceptor agonists (Kable et al. 2000).

In addition to these presynaptic actions, postsynaptic changes have been described. Glycinergic membrane currents in rat sacral commissural neurons (lamina X) are potentiated by norepinephrine acting on $\alpha 2$ adrenoceptors (Nabekura et al. 1999). This potentiation is mediated by a decrease in cAMP and inhibition of PKA, and is prevented by pretreatment with pertussis toxin, clearly indicating that potentiation is due to a reversal of PKA-mediated inhibition. Although GlyR$\alpha 3$ is not expressed in lamina X of the spinal cord (at least not in adult mice), these features are very reminiscent of the inhibition of GlyR$\alpha 3$ by prostaglandin E2 discussed above. Norepinephrine might thus act as an endogenous functional antagonist counteracting prostaglandin E2-mediated central pain sensitization.

The actions of serotonin are more complex. Activation of G-protein-coupled 5-HT receptors by serotonin potentiates $GABA_A$ receptor-mediated responses in lamina X (Xu et al. 1998) and superficial dorsal horn (Li et al. 2000) neurons through activation of protein kinase C. A similar effect also occurs in glycine receptors (Xu et al. 1996). In addition, 5-HT1D receptors reduce the occurrence of spontaneous glutamatergic excitatory postsynaptic currents (EPSCs; Travagli and Williams 1996). These effects probably account for the antinociceptive effects of descending serotoninergic fiber tracts. Activation of ionotropic 5-HT3 receptors has been reported both to inhibit (Khasabov et al. 1999) and to increase (Hori et al. 1996) glutamate release from primary afferents. Apart from its acute effects on inhibitory and excitatory synaptic transmission, serotonin also appears to trigger longer-lasting changes. The incidence of long-term depression after conditioning stimulation was increased in deep dorsal horn neurons, although baseline excitatory synaptic transmission was reduced (Garraway and Hochman 2001). In addition, serotonin induces a long-lasting facilitation of evoked action potentials and spontaneous glutamate release (Hori et al. 1996).

Cannabinoid receptors

Since the discovery of the two cannabinoid receptors CB-1 and CB-2 and of their endogenous activator anandamide, intense research efforts have been made to address their physiological role and to probe their potentials as drug targets. Besides central spasticity (such as in patients suffering from multiple sclerosis) and anxiety, pain has been the focus of research right from the beginning of cannabinoid research. Activation of CB-1 receptors by anandamide, arachidonyl-2-chloroethylamide, and WIN-55,212–2 inhibits glutamatergic transmission between primary nociceptive afferents and second-order neurons in the superficial dorsal horn (Harris et al. 2000; Kelly and Chapman 2001; Morisset and Urban 2001). The functional consequences of CB1 receptor activation in the spinal cord are thus somewhat reminiscent of those of opioid receptors, with which CB-1 receptors are coexpressed in the dorsal horn (Salio et al. 2001). It should, however, be noted that the physiological function of endogenous anandamide in the dorsal horn is at present difficult to judge because anandamide also possesses agonistic activity at pronociceptive capsaicin receptors (Tognetto et al. 2001).

Nicotine receptors

Activation of nicotinic receptors evokes profound analgesia. Unfortunately, nicotinic agonists exert toxic effects at doses only slightly higher than those needed for analgesia (for a review, see Flores 2000). In the dorsal horn, nicotinic receptors are expressed on presynaptic terminals of inhibitory neurons (Takeda et al. 2003) and on the spinal terminals of serotoninergic neurons descending from the raphe magnus (Cordero-Erausquin and Changeux 2001). In both cases, nicotinic receptor activation facilitates transmitter release and enhances nociceptive inhibition. The facilitating effect on glycine and GABA release persists in the presence of dihydro-beta-erythroidine and methyllycaconitine, indicating that it is not mediated through $\alpha4\beta2$- or $\alpha7$-containing nicotine receptors (Takeda et al. 2003, but see also Kiyosawa et al. 2001). Expression analysis employing single-cell reverse transcription-polymerase chain reaction suggests a dominant role of $\alpha4\alpha6\beta2$ nicotine receptors on inhibitory neurons in the dorsal horn (Cordero-Erausquin et al. 2004), making this receptor isoform a preferred target for possible nicotinergic analgesics.

Purines

Adenosine triphosphate (ATP) facilitates glycine (Rhee et al. 2000; Jang et al. 2001), GABA (Hugel and Schlichter 2000), and glutamate (Gu and MacDermott 1997; Li et al. 1998) release through the activation of P2X receptors in the spinal cord. Because ATP is coreleased with GABA from many dorsal horn interneurons (Jo and Schlichter 1999), it is likely that ATP plays a role as an endogenous regulator of dorsal horn neurotransmission. It has also been proposed that ATP functions as a fast synaptic transmitter in primary afferents (Bardoni et al. 1997), but the majority of reports favor the function of a modulator rather than of a principle transmitter. Interestingly, ADP and adenosine, which originate from the fast breakdown of ATP in the extracellular space, exert opposite effects. Activation of G-protein-coupled P2Y receptors by ADP inhibits N-type Ca^{2+} currents in presumed nociceptive rat dorsal root ganglion neurons (Gerevich et al. 2004). Adenosine inhibits glycine and GABA release (Yang et al. 2004; Hugel and Schlichter 2003), but at the same time it also reduces synaptic transmission between primary afferent C and Aδ fibers via a presynaptic mechanism (Lao et al. 2004). Both effects are consistent with the dense expression of A1 adenosine receptors in the inner part of lamina I (Schulte et al. 2003). Inhibition of excitatory transmission between primary nociceptive afferents and spinal neurons probably underlies the well-documented spinal antinociceptive effects of adenosine.

GABA$_B$ receptors

In behavioral tests, baclofen and CGP-35024, two GABA$_B$ receptor agonists, evoke antinociception after systemic or intrathecal injection (Patel et al. 2001). Mice lacking one of the GABA$_B$ receptor subunits exhibit increased acute pain sensitivity, suggesting that GABA$_B$ receptor activation contributes to tonic pain control (Schuler et al. 2001; Gassmann et al. 2004). On the cellular level, baclofen inhibits transmitter release from primary afferent C fibers and to a lesser extent from Aβ fibers (Ataka et al. 2000), but also from inhibitory nerve terminals in the dorsal horn (Iyadomi et al. 2000).

Presynaptic Ca^{2+} channels

Many of the antinociceptive neuromodulators discussed above exert their effects through inhibiting transmitter release from primary afferent nerve fibers. In most cases this inhibition probably occurs through a direct G-protein-mediated inhibition of voltage-gated Ca^{2+} channels in the membrane of presynaptic terminals. Glutamate release from primary nociceptive afferents is triggered by voltage-gated Ca^{2+} channels of the Cav2 family, which comprises P/Q-type (Cav2.1), N-type (Cav2.2), and R-type (Cav2.3) Ca^{2+} channels. All three channel types can be blocked with high specificity by peptide toxins derived from marine snails and spiders. In particular, blockade of N-type channels with N-type Ca^{2+} channel toxins has proved to provide efficient analgesia after intrathecal injection (e.g., Bowersox et al. 1996). Accordingly, mice lacking the pore-forming subunit α1B of N-type Ca^{2+} channels or the accessory β3 subunit, which preferentially associates with N-type Ca^{2+} channels, exhibit reduced nociceptive responses in different tests of nociception (Saegusa et al. 2001; Murakami et al. 2002). These observations have already led to the use of other N-type channel toxins (ziconotide, also SNX-111 or ω-conotoxin MVIIC, and AM336 or ω-conotoxin CVID) for treating otherwise intractable pain (reviewed by Miljanich 2004). A clear advantage of direct Ca^{2+} channel blockade over modulation by, for instance, opioids is the lack of tolerance development; a disadvantage is certainly that the peptide toxins available at present do not penetrate the blood–brain barrier and therefore have to be injected via chronically implanted intrathecal catheters.

The contribution of other Ca^{2+} channel subtypes of the Cav2 family to spinal nociceptive transmission is less clear. Blockade of P/Q-type Ca^{2+} channels with ω-agatoxin IVA is analgesic in arthritic but not in normal rats (Nebe et al. 1997). Similarly, mice lacking the pore-forming subunit α 1E of R-type (Cav2.3) Ca^{2+} channels also show normal responses to acute pain, but reduced inflammatory pain (Saegusa et al. 2000).

A Ca^{2+} channel targeting drug that is already very successful on the market as a centrally-acting analgesic is gabapentin. Gabapentin, 1-(aminomethyl) cyclohexane acetic acid, was originally developed as an anticonvulsant, but in the meantime it has been widely used for treating neuropathic pain (Bennett and Simpson 2004). Despite numerous studies that have demonstrated clear analgesic activity in both clinical and preclinical tests (Mao and Chen 2000), the mode of action of gabapentin is still ill-defined. In contrast to its structural similarity to GABA, it does not bind to ionotropic $GABA_A$ or G-protein coupled $GABA_B$ receptors. The only known high-affinity binding site is the accessory Ca^{2+} channel subunit α2δ, which exists in four isoforms (Marais et al. 2001), of which only two (α2δ-1 and α2δ-2) bind gabapentin (Klugbauer et al. 2003; Marais et al. 2001). Recent data using mice carrying a mutation in the α2δ-1 isoform suggest that binding of gabapentin to this subunit is indeed responsible for gabapentin's antinociceptive (and anticonvulsive) effects (Taylor 2004). The importance of gabapentin binding to α2δ–1 is also supported by increased expression of this subunit in dorsal root ganglion neurons during neuropathic pain (Luo et al. 2001), which correlates with the antinociceptive effect of gabapentin (Z.D. Luo et al. 2002b) and with its spinal site of action (Shimoyama et al. 2000; Jun and Yaksh 1998; Kaneko et al. 2000). However, the effects that induction of α2δ has on Ca^{2+} functions and the ways that Ca^{2+} channel function is affected by the binding of gabapentin remain elusive. Several studies, including one in the spinal cord (Bayer et al. 2004), suggest that binding of gabapentin to Ca^{2+} channels impairs neurotransmitter release mediated by P/Q-type Ca^{2+} channels (Meder and Dooley 2000; Dooley et al. 2002; Fink et al. 2000, 2002). As a consequence of Ca^{2+} channel inhibition, gabapentin reduces the release of neurotransmitters in the dorsal horn without discriminating between excitatory and inhibitory neurotransmitters (Bayer et al. 2004). It should, however, also be noted here that the preferential action on Ca^{2+} channel types may vary between different CNS areas (Brown and Randall 2005), as association of the type-determining Ca^{2+} channel subunits with accessory subunits differs considerably within the CNS. It is therefore not surprising that a number of conflicting reports have been published (Gu and Huang 2002, Moore et al. 2002b). At present, how this inhibition relates to the analgesic action of gabapentin is unknown.

Conclusions

Significant progress has been made during the last decade in our understanding of the molecular events leading to activity-dependent and inflammation-induced pain sensitization. In particular, the use of genetically engineered mice in pain research has helped to unravel a number of new potential targets for the development of novel analgesics. These range from specific prostaglandin receptor antagonists and blockers of prostaglandin synthases to allosteric modulators of inhibitory neurotransmitter receptors and protein kinase inhibitors.

Unlike inflammatory pain, neuropathic pain is still a scientific mystery and a great therapeutic challenge. Although several important findings have been made recently, a clear concept of how peripheral nerve injury affects the excitability of central nociceptive neurons is only gradually emerging.

References

Ahmadi S, Kotalla C, Gühring H, Takeshima H, Pahl A, Zeilhofer HU (2001) Modulation of synaptic transmission by nociceptin/orphanin FQ and nocistatin in the spinal cord dorsal horn of mutant mice lacking the nociceptin/orphanin FQ receptor. Mol Pharmacol 59:612–618

Ahmadi S, Lippross S, Neuhuber WL, Zeilhofer HU (2002) PGE$_2$ selectively blocks inhibitory glycinergic neurotransmission onto rat superficial dorsal horn neurons. Nat Neurosci 5:34–40

Ahmadi S, Muth-Selbach U, Lauterbach A, Lipfert P, Neuhuber WL, Zeilhofer HU (2003) Facilitation of spinal NMDA receptor currents by spillover of synaptically released glycine. Science 300:2094–2097

Albuquerque C, Lee CJ, Jackson AC, MacDermott AB (1999) Subpopulations of GABAergic and non-GABAergic rat dorsal horn neurons express Ca^{2+}-permeable AMPA receptors. Eur J Neurosci 11:2758–2766

Antal M, Petko M, Polgar E, Heizmann CW, Storm-Mathisen J (1996) Direct evidence of an extensive GABAergic innervation of the spinal dorsal horn by fibres descending from the rostral ventromedial medulla. Neuroscience 73:509–518

Ataka T, Kumamoto E, Shimoji K, Yoshimura M (2000) Baclofen inhibits more effectively C-afferent than Adelta-afferent glutamatergic transmission in substantia gelatinosa neurons of adult rat spinal cord slices. Pain 86:273–282

Azkue JJ, Liu XG, Zimmermann M, Sandkuhler J (2003) Induction of long-term potentiation of C fibre-evoked spinal field potentials requires recruitment of group I, but not group II/III metabotropic glutamate receptors. Pain 106:373–379

Baba H, Doubell TP, Moore KA, Woolf CJ (2000a) Silent NMDA receptor-mediated synapses are developmentally regulated in the dorsal horn of the rat spinal cord. J Neurophysiol 83:955–962

Baba H, Goldstein PA, Okamoto M, Kohno T, Ataka T, Yoshimura M, Shimoji K (2000b) Norepinephrine facilitates inhibitory transmission in substantia gelatinosa of adult rat spinal cord. II. Effects on somato-dendritic sites of GABAergic neurons. Anesthesiology 92:485–492

Baba H, Shimoji K, Yoshimura M (2000c) Norepinephrine facilitates inhibitory transmission in substantia gelatinosa of adult rat spinal cord. I. Effects on axon terminals of GABAergic and glycinergic neurons. Anesthesiology 92:473–484

Baba H, Kohno T, Moore KA, Woolf CJ (2001) Direct activation of rat spinal dorsal horn neurons by prostaglandin E2. J Neurosci 21:1750–1756

Bär KJ, Natura G, Telleria-Diaz A, Teschner P, Vogel R, Vasquez E, Schaible HG, Ebersberger A (2004) Changes in the effect of spinal prostaglandin E2 during inflammation: prostaglandin E (EP1-EP4) receptors in spinal nociceptive processing of input from the normal or inflamed knee joint. J Neurosci 24:642–651

Bardoni R, Goldstein PA, Lee CJ, Gu JG, MacDermott AB (1997) ATP P2X receptors mediate fast synaptic transmission in the dorsal horn of the rat spinal cord. J Neurosci 17:5297–5304

Bardoni R, Torsney C, Tong CK, Prandini M, MacDermott AB (2004) Presynaptic NMDA receptors modulate glutamate release from primary sensory neurons in rat spinal cord dorsal horn. J Neurosci 24:2774–2781

Barria A, Muller D, Derkach V, Griffith LC, Soderling TR (1997) Regulatory phosphorylation of AMPA-type glutamate receptors by CaM-KII during long-term potentiation. Science 276:2042–2045

Basbaum AI, Fields HL (1984) Endogenous pain control systems: brainstem spinal pathways and endorphin circuitry. Annu Rev Neurosci 7:309–338

Bayer K, Ahmadi S, Zeilhofer HU (2004) Gabapentin may inhibit synaptic transmission in the mouse spinal cord dorsal horn through a preferential block of P/Q-type Ca^{2+} channels. Neuropharmacology 46:743–749

Beiche F, Scheuerer S, Brune K, Geisslinger G, Goppelt-Struebe M (1996) Up-regulation of cyclooxygenase-2 mRNA in the rat spinal cord following peripheral inflammation. FEBS Lett 390:165–169

Bennett MI, Simpson KH (2004) Gabapentin in the treatment of neuropathic pain. Palliat Med 18:5–11

Berger AJ, Dieudonne S, Ascher P (1998) Glycine uptake governs glycine site occupancy at NMDA receptors of excitatory synapses. J Neurophysiol 80:3336–3340

Bergeron R, Meyer TM, Coyle JT, Greene RW (1998) Modulation of N-methyl-D-aspartate receptor function by glycine transport. Proc Natl Acad Sci USA 95:15730–15734

Bowersox SS, Gadbois T, Singh T, Pettus M, Wang YX, Luther RR (1996) Selective N-type neuronal voltage-sensitive calcium channel blocker, SNX-111, produces spinal antinociception in rat models of acute, persistent and neuropathic pain. J Pharmacol Exp Ther 279:1243–1249

Breitinger HG, Becker CM (1998) The inhibitory glycine receptor: prospects for a therapeutic orphan? Curr Pharm Des 4:315–334

Broom DC, Samad TA, Kohno T, Tegeder I, Geisslinger G, Woolf CJ (2004) Cyclooxygenase 2 expression in the spared nerve injury model of neuropathic pain. Neuroscience 124:891–900

Brown JT, Randall A (2005) Gabapentin fails to alter P/Q-type Ca^{2+} channel-mediated synaptic transmission in the hippocampus. Synapse 55:262–269

Cervero F, Laird JM (1996) Mechanisms of allodynia: interactions between sensitive mechanoreceptors and nociceptors. Neuroreport 7:526–528

Chen J, Heinke B, Sandkühler J (2000) Activation of group I metabotropic glutamate receptors induces long-term depression at sensory synapses in superficial spinal dorsal horn. Neuropharmacology 39:2231–2243

Chen L, Huang LY (1992) Protein kinase C reduces Mg^{2+} block of NMDA-receptor channels as a mechanism of modulation. Nature 356:521–523

Chery N, de Koninck Y (1999) Junctional versus extrajunctional glycine and $GABA_A$ receptor-mediated IPSCs in identified lamina I neurons of the adult rat spinal cord. J Neurosci 19:7342–7355

Chery N, de Koninck Y (2000) $GABA_B$ receptors are the first target of released GABA at lamina I inhibitory synapses in the adult rat spinal cord. J Neurophysiol 84:1006–1011

Chizh BA, Headley PM, Tzschentke TM (2001) NMDA receptor antagonists as analgesics: focus on the NR2B subtype. Trends Pharmacol Sci 22:636–642

Coderre TJ, Melzack R (1987) Cutaneous hyperalgesia: contributions of the peripheral and central nervous systems to the increase in pain sensitivity after injury. Brain Res 404:95–106

Cook AJ, Woolf CJ, Wall PD, McMahon SB (1987) Dynamic receptive field plasticity in rat spinal cord dorsal horn following C-primary afferent input. Nature 325:151–153

Cordero-Erausquin M, Changeux JP (2001) Tonic nicotinic modulation of serotoninergic transmission in the spinal cord. Proc Natl Acad Sci USA 98:2803–2807

Cordero-Erausquin M, Pons S, Faure P, Changeux JP (2004) Nicotine differentially activates inhibitory and excitatory neurons in the dorsal spinal cord. Pain 109:308–318

Coull JA, Boudreau D, Bachand K, Prescott SA, Nault F, Sik A, de Koninck P, de Koninck Y (2003) Trans-synaptic shift in anion gradient in spinal lamina I neurons as a mechanism of neuropathic pain. Nature 424:938–942

Depner UB, Reinscheid RK, Takeshima H, Brune K, Zeilhofer HU (2003) Normal sensitivity of acute pain, but increased inflammatory hyperalgesia in mice lacking the nociceptin precursor polypeptide or the nociceptin receptor. Eur J Neurosci 17:2381–2387

Derkach V, Barria A, Soderling TR (1999) Ca^{2+}/calmodulin-kinase II enhances channel conductance of alpha-amino-3-hydroxy-5-methyl-4-isoxazolepropionate type glutamate receptors. Proc Natl Acad Sci USA 96:3269–3274

Dooley DJ, Donovan CM, Meder WP, Whetzel SZ (2002) Preferential action of gabapentin and pregabalin at P/Q-type voltage-sensitive calcium channels: inhibition of K^+-evoked [3H]-norepinephrine release from rat neocortical slices. Synapse 45:171–190

Durand GM, Kovalchuk Y, Konnerth A (1996) Long-term potentiation and functional synapse induction in developing hippocampus. Nature 381:71–75

Engelman HS, Allen TB, MacDermott AB (1999) The distribution of neurons expressing calcium-permeable AMPA receptors in the superficial laminae of the spinal cord dorsal horn. J Neurosci 19:2081–2089

Erb K, Liebel JT, Tegeder I, Zeilhofer HU, Brune K, Geisslinger G (1997) Spinally delivered nociceptin/orphanin FQ reduces flinching behaviour in the rat formalin test. Neuroreport 8:1967–1970

Esteban JA, Shi SH, Wilson C, Nuriya M, Huganir RL, Malinow R (2003) PKA phosphorylation of AMPA receptor subunits controls synaptic trafficking underlying plasticity. Nat Neurosci 6:136–143

Fang L, Wu J, Zhang X, Lin Q, Willis WD (2003) Increased phosphorylation of the GluR1 subunit of spinal cord alpha-amino-3-hydroxy-5-methyl-4-isoxazole propionate receptor in rats following intradermal injection of capsaicin. Neuroscience 122:237–245

Fink K, Meder W, Dooley DJ, Göthert M (2000) Inhibition of neuronal Ca^{2+} influx by gabapentin and subsequent reduction of neurotransmitter release from rat neocortical slices. Br J Pharmacol 130:900–906

Fink K, Dooley DJ, Meder WP, Suman-Chauhan N, Duffy S, Clusmann H, Göthert M (2002) Inhibition of neuronal Ca^{2+} influx by gabapentin and pregabalin in the human neocortex. Neuropharmacology 42:229–236

Flores CM (2000) The promise and pitfalls of a nicotinic cholinergic approach to pain management. Pain 88:1–6

Gabernet L, Pauly-Evers M, Schwerdel C, Lentz M, Bluethmann H, Vogt K, Alberati D, Möhler H, Boison D (2005) Enhancement of the NMDA receptor function by reduction of glycine transporter-1 expression. Neurosci Lett 373:79–84

Galan A, Laird JM, Cervero F (2004) In vivo recruitment by painful stimuli of AMPA receptor subunits to the plasma membrane of spinal cord neurons. Pain 112:315–323

Garraway SM, Hochman S (2001) Serotonin increases the incidence of primary afferent-evoked long-term depression in rat deep dorsal horn neurons. J Neurophysiol 85:1864–1872

Gassmann M, Shaban H, Vigot R, Sansig G, Haller C, Barbieri S, Humeau Y, Schuler V, Muller M, Kinzel B, Klebs K, Schmutz M, Froestl W, Heid J, Kelly PH, Gentry C, Jaton AL, Van der Putten H, Mombereau C, Lecourtier L, Mosbacher J, Cryan JF, Fritschy JM, Luthi A, Kaupmann K, Bettler B (2004) Redistribution of GABAB$_1$ protein and atypical GABAB responses in GABAB$_2$-deficient mice. J Neurosci 24:6086–6097

Gerevich Z, Borvendeg SJ, Schroder W, Franke H, Wirkner K, Norenberg W, Furst S, Gillen C, Illes P (2004) Inhibition of N-type voltage-activated calcium channels in rat dorsal root ganglion neurons by P2Y receptors is a possible mechanism of ADP-induced analgesia. J Neurosci 24:797–807

Goodchild CS, Serrao JM (1987) Intrathecal midazolam in the rat: evidence for spinally-mediated analgesia. Br J Anaesth 59:1563–1570

Grisel JE, Mogil JS, Belknap JK, Grandy DK (1996) Orphanin FQ acts as a supraspinal, but not a spinal, anti-opioid peptide. Neuroreport 7:2125–2129

Grudt TJ, Henderson G (1998) Glycine and GABA$_A$ receptor-mediated synaptic transmission in rat substantia gelatinosa: inhibition by μ-opioid and GABAB agonists. J Physiol 507:473–483

Gu JG, MacDermott AB (1997) Activation of ATP P2X receptors elicits glutamate release from sensory neuron synapses. Nature 389:749–753

Gu JG, Albuquerque C, Lee CJ, MacDermott AB (1996) Synaptic strengthening through activation of Ca^{2+}-permeable AMPA receptors. Nature 381:793–796

Gu Y, Huang LY (2002) Gabapentin potentiates N-methyl-D-aspartate receptor mediated currents at GABAergic dorsal horn neurons. Neurosci Lett 324:177–180

Guay J, Bateman K, Gordon R, Mancini J, Riendeau D (2004) Carrageenan-induced paw edema in rat elicits a predominant prostaglandin E2 (PGE2) response in the central nervous system associated with the induction of microsomal PGE2 synthase-1. J Biol Chem 279:24866–24872

Guo W, Wei F, Zou S, Robbins MT, Sugiyo S, Ikeda T, Tu JC, Worley PF, Dubner R, Ren K (2004) Group I metabotropic glutamate receptor NMDA receptor coupling and signaling cascade mediate spinal dorsal horn NMDA receptor 2B tyrosine phosphorylation associated with inflammatory hyperalgesia. J Neurosci 24:9161–9173

Harris J, Drew LJ, Chapman V (2000) Spinal anandamide inhibits nociceptive transmission via cannabinoid receptor activation in vivo. Neuroreport 11:2817–2819

Hartmann B, Ahmadi S, Heppenstall PA, Lewin GR, Schott C, Borchardt T, Seeburg PH, Zeilhofer HU, Sprengel R, Kuner R (2004) The AMPA receptor subunits GluR-A and GluR-B reciprocally modulate spinal synaptic plasticity and inflammatory pain. Neuron 44:637–650

Harvey RJ, Depner UB, Wässle H, Ahmadi S, Heindl C, Reinold H, Smart TG, Harvey K, Schütz B, Abo-Salem OM, Zimmer A, Poisbeau P, Welzl H, Wolfer DP, Betz H, Zeilhofer HU, Müller U (2004) GlyR α3: an essential target for spinal PGE$_2$-mediated inflammatory pain sensitization. Science 304:884–887

Hebb DO (1966) A textbook of psychology. Saunders, Philadelphia

Hori Y, Endo K, Takahashi T (1996) Long-lasting synaptic facilitation induced by serotonin in superficial dorsal horn neurones of the rat spinal cord. J Physiol 492:867–876

Hugel S, Schlichter R (2000) Presynaptic P2X receptors facilitate inhibitory GABAergic transmission between cultured rat spinal cord dorsal horn neurons. J Neurosci 20:2121–2130

Hugel S, Schlichter R (2003) Convergent control of synaptic GABA release from rat dorsal horn neurones by adenosine and GABA autoreceptors. J Physiol 551:479–489

Hunter JC, Fontana DJ, Hedley LR, Jasper JR, Lewis R, Link RE, Secchi R, Sutton J, Eglen RM (1997) Assessment of the role of alpha2-adrenoceptor subtypes in the antinociceptive, sedative and hypothermic action of dexmedetomidine in transgenic mice. Br J Pharmacol 122:1339–1344

Hwang SJ, Pagliardini S, Rustioni A, Valtschanoff JG (2001) Presynaptic kainate receptors in primary afferents to the superficial laminae of the rat spinal cord. J Comp Neurol 436:275–289

Ikeda H, Heinke B, Ruschewey R, Sandkühler J (2003) Synaptic plasticity in spinal lamina I projection neurons that mediate hyperalgesia. Science 299:1237–1240

Iyadomi M, Iyadomi I, Kumamoto E, Tomokuni K, Yoshimura M (2000) Presynaptic inhibition by baclofen of miniature EPSCs and IPSCs in substantia gelatinosa neurons of the adult rat spinal dorsal horn. Pain 85:385–393

Jang IS, Rhee JS, Kubota H, Akaike N (2001) Developmental changes in P2X purinoceptors on glycinergic presynaptic nerve terminals projecting to rat substantia gelatinosa neurones. J Physiol 536:505–519

Jasmin L, Wu MV, Ohara PT (2004) GABA puts a stop to pain. Curr Drug Targets CNS Neurol Disord 3:487–505

Ji RR, Kohno T, Moore KA, Woolf CJ (2003) Central sensitization and LTP: do pain and memory share similar mechanisms? Trends Neurosci 26:696–705

Jo YH, Schlichter R (1999) Synaptic corelease of ATP and GABA in cultured spinal neurons. Nat Neurosci 2:241–245

Johnson JW, Ascher P (1987) Glycine potentiates the NMDA response in cultured mouse brain neurons. Nature 325:529–531

Jun JH, Yaksh TL (1998) The effect of intrathecal gabapentin and 3-isobutyl gamma-aminobutyric acid on the hyperalgesia observed after thermal injury in the rat. Anesth Analg 86:348–354

Kable JW, Murrin LC, Bylund DB (2000) In vivo gene modification elucidates subtype-specific functions of α_2-adrenergic receptors. J Pharmacol Exp Ther 293:1–7

Kamei D, Yamakawa K, Takegoshi Y, Mikami-Nakanishi M, Nakatani Y, Oh-Ishi S, Yasui H, Azuma Y, Hirasawa N, Ohuchi K, Kawaguchi H, Ishikawa Y, Ishii T, Uematsu S, Akira S, Murakami M, Kudo I (2004) Reduced pain hypersensitivity and inflammation in mice lacking microsomal prostaglandin E synthase-1. J Biol Chem 279:33684–33695

Kaneko M, Mestre C, Sanchez EH, Hammond DL (2000) Intrathecally administered gabapentin inhibits formalin-evoked nociception and Fos-like immunoreactivity in the spinal cord of the rat. J Pharmacol Exp Ther 292:743–751

Kawasaki Y, Kumamoto E, Furue H, Yoshimura M (2003) $\alpha2$ adrenoceptor-mediated presynaptic inhibition of primary afferent glutamatergic transmission in rat substantia gelatinosa neurons. Anesthesiology 98:682–689

Keller AF, Coull JA, Chery N, Poisbeau P, de Koninck Y (2001) Region-specific developmental specialization of GABA-glycine cosynapses in laminas I-II of the rat spinal dorsal horn. J Neurosci 21:7871–7880

Kelly S, Chapman V (2001) Selective cannabinoid CB1 receptor activation inhibits spinal nociceptive transmission in vivo. J Neurophysiol 86:3061–3064

Kerchner GA, Wang GD, Qiu CS, Huettner JE, Zhuo M (2001a) Direct presynaptic regulation of GABA/glycine release by kainate receptors in the dorsal horn: an ionotropic mechanism. Neuron 32:477–488

Kerchner GA, Wilding TJ, Li P, Zhuo M, Huettner JE (2001b) Presynaptic kainate receptors regulate spinal sensory transmission. J Neurosci 21:59–66

Khasabov SG, Lopez-Garcia JA, Asghar AU, King AE (1999) Modulation of afferent-evoked neurotransmission by 5-HT3 receptors in young rat dorsal horn neurones in vitro: a putative mechanism of 5-HT3 induced anti-nociception. Br J Pharmacol 127:843–852

Khasabov SG, Rogers SD, Ghilardi JR, Peters CM, Mantyh PW, Simone DA (2002) Spinal neurons that possess the substance P receptor are required for the development of central sensitization. J Neurosci 22:9086–9098

Kiyosawa A, Katsurabayashi S, Akaike N, Pang ZP, Akaike N (2001) Nicotine facilitates glycine release in the rat spinal dorsal horn. J Physiol 536:101–110

Kleckner NW, Dingledine R (1988) Requirement for glycine in activation of NMDA-receptors expressed in Xenopus oocytes. Science 241:835–837

Klein T, Magerl W, Hopf HC, Sandkühler J, Treede RD (2004) Perceptual correlates of nociceptive long-term potentiation and long-term depression in humans. J Neurosci 24:964–971

Klugbauer N, Marais E, Hofmann F (2003) Calcium channel $\alpha2\delta$ subunits: differential expression, function, and drug binding. J Bioenerg Biomembranes 35:639–647

Kohno T, Kumamoto E, Higashi H, Shimoji K, Yoshimura M (1999) Actions of opioids on excitatory and inhibitory transmission in substantia gelatinosa of adult rat spinal cord. J Physiol 518:803–813

Krogsgaard-Larsen P, Frolund B, Liljefors T, Ebert B (2004) GABA$_A$ agonists and partial agonists: THIP (Gaboxadol) as a non-opioid analgesic and a novel type of hypnotic. Biochem Pharmacol 68:1573–1580

Lakhlani PP, Macmillan LB, Guo TZ, McCool BA, Lovinger DM, Maze M, Limbird LE (1997) Substitution of a mutant $\alpha2a$-adrenergic receptor via "hit and run" gene targeting reveals the role of this subtype in sedative, analgesic, and anesthetic-sparing responses in vivo. Proc Natl Acad Sci USA 94:9950–9955

Lao LJ, Kawasaki Y, Yang K, Fujita T, Kumamoto E (2004) Modulation by adenosine of Aδ and C primary-afferent glutamatergic transmission in adult rat substantia gelatinosa neurons. Neuroscience 125:221–231

Laube B, Maksay G, Schemm R, Betz H (2002) Modulation of glycine receptor function: a novel approach for therapeutic intervention at inhibitory synapses? Trends Pharmacol Sci 23:519–527

Lee TL, Fung FM, Chen FG, Chou N, Okuda-Ashitaka E, Ito S, Nishiuchi Y, Kimura T, Tachibana S (1999) Identification of human, rat and mouse nocistatin in brain and human nocistatin in brain and human cerebrospinal fluid. Neuroreport 10:1537–1541

Legendre P (2001) The glycinergic inhibitory synapse. Cell Mol Life Sci 58:760–793

Li H, Lang B, Kang JF, Li YQ (2000) Serotonin potentiates the response of neurons of the superficial laminae of the rat spinal dorsal horn to gamma-aminobutyric acid. Brain Res Bull 52:559–565

Li P, Zhuo M (1998) Silent glutamatergic synapses and nociception in mammalian spinal cord. Nature 393:695–698

Li P, Calejesan AA, Zhuo M (1998) ATP P2x receptors and sensory synaptic transmission between primary afferent fibers and spinal dorsal horn neurons in rats. J Neurophysiol 80:3356–3360

Li P, Wilding TJ, Kim SJ, Calejesan AA, Huettner JE, Zhuo M (1999) Kainate-receptor-mediated sensory synaptic transmission in mammalian spinal cord. Nature 397:161–164

Liebel JT, Swandulla D, Zeilhofer HU (1997) Modulation of excitatory synaptic transmission by nociceptin in superficial dorsal horn neurones of the neonatal rat spinal cord. Br J Pharmacol 121:425–432

Light AR, Willcockson HH (1999) Spinal laminae I-II neurons in rat recorded in vivo in whole cell, tight seal configuration: properties and opioid responses. J Neurophysiol 82:3316–3326

Liu XG, Sandkühler J (1995) Long-term potentiation of C-fiber-evoked potentials in the rat spinal dorsal horn is prevented by spinal N-methyl-D-aspartic acid receptor blockage. Neurosci Lett 191:43–46

Liu H, Wang H, Sheng M, Jan LY, Jan YN, Basbaum AI (1994) Evidence for presynaptic N-methyl-D-aspartate autoreceptors in the spinal cord dorsal horn. Proc Natl Acad Sci USA 91:8383–8387

Liu H, Mantyh PW, Basbaum AI (1997) NMDA-receptor regulation of substance P release from primary afferent nociceptors. Nature 386:721–724

Liu XG, Morton CR, Azkue JJ, Zimmermann M, Sandkühler J (1998) Long-term depression of C-fibre-evoked spinal field potentials by stimulation of primary afferent A delta-fibres in the adult rat. Eur J Neurosci 10:3069–3075

Löw K, Crestani F, Keist R, Benke D, Brünig I, Benson JA, Fritschy JM, Rülicke T, Bluethmann H, Möhler H, Rudolph U (2000) Molecular and neuronal substrate for the selective attenuation of anxiety. Science 290:131–134

Lu CR, Hwang SJ, Phend KD, Rustioni A, Valtschanoff JG (2002) Primary afferent terminals in spinal cord express presynaptic AMPA receptors. J Neurosci 22:9522–9529

Luo ZD, Chaplan SR, Higuera ES, Sorkin LS, Stauderman KA, Williams ME, Yaksh TL (2001) Upregulation of dorsal root ganglion $\alpha2\delta$ calcium channel subunit and its correlation with allodynia in spinal nerve-injured rats. J Neurosci 21:1868–1875

Luo C, Kumamoto E, Furue H, Chen J, Yoshimura M (2002a) Nociceptin inhibits excitatory but not inhibitory transmission to substantia gelatinosa neurones of adult rat spinal cord. Neuroscience 109:349–358

Luo ZD, Calcutt NA, Higuera ES, Valder CR, Song YH, Svensson CI, Myers RR (2002b) Injury type-specific calcium channel alpha 2 delta-1 subunit up-regulation in rat neuropathic pain models correlates with an-tiallodynic effects of gabapentin. J Pharmacol Exp Ther 303:1199–1205

Lynch JW (2004) Molecular structure and function of the glycine receptor chloride channel. Physiol Rev 84:1051–1095

Mack V, Burnashev N, Kaiser KM, Rozov A, Jensen V, Hvalby O, Seeburg PH, Sakmann B, Sprengel R (2001) Conditional restoration of hippocampal synaptic potentiation in Glur-A-deficient mice. Science 292:2501–2504

Malmberg AB, Brandon EP, Idzerda RL, Liu H, McKnight GS, Basbaum AI (1997) Diminished inflammation and nociceptive pain with preservation of neuropathic pain in mice with a targeted mutation of the type I regulatory subunit of cAMP-dependent protein kinase. J Neurosci 17:7462–7470

Mantyh PW, Rogers SD, Honore P, Allen BJ, Ghilardi JR, Li J, Daughters RS, Lappi DA, Wiley RG, Simone DA (1997) Inhibition of hyperalgesia by ablation of lamina I spinal neurons expressing the substance P receptor. Science 278:275–279

Mao J, Chen LL (2000) Gabapentin in pain management. Anesth Analg 91:680–687

Marais E, Klugbauer N, Hofmann F (2001) Calcium channel $\alpha2\delta$ subunits-structure and gabapentin binding. Mol Pharmacol 59:1243–1248

Matthes HW, Maldonado R, Simonin F, Valverde O, Slowe S, Kitchen I, Befort K, Dierich A, Le Meur M, Dolle P, Tzavara E, Hanoune J, Roques BP, Kieffer BL (1996) Loss of morphine-induced analgesia, reward effect and withdrawal symptoms in mice lacking the mu-opioid-receptor gene. Nature 383:819–823

McKernan RM, Rosahl TW, Reynolds DS, Sur C, Wafford K, Atack JR, Farrar S, Myers J, Cook G, Ferris P, Garrett L, Bristow L, Marshall G, Macaulay A, Brown N, Howell O, Moore KW, Carling RW, Street LJ, Castro JL, Ragan CI, Dawson GR, Whiting PJ (2000) Sedative but not anxiolytic properties of benzodiazepines are mediated by the GABA$_A$ receptor alpha1 subtype. Nat Neurosci 3:587–592

Meder WP, Dooley DJ (2000) Modulation of K^+-induced synaptosomal calcium influx by gabapentin. Brain Res 875:157–159

Mendell LM (1966) Physiological properties of unmyelinated fiber projections to the spinal cord. Exp Neurol 16:316–332

Meunier JC, Mollereau C, Toll L, Suaudeau C, Moisand C, Alvinerie P, Butour JL, Guillemot JC, Ferrara P, Monsarrat B, Marzarguil H, Vassart G, Parmentier M, Costentin J (1995) Isolation and structure of the endogenous agonist of opioid receptor-like ORL1 receptor. Nature 377:532–535

Miljanich GP (2004) Ziconotide: neuronal calcium channel blocker for treating severe chronic pain. Curr Med Chem 11:3029–3040

Minami T, Okuda-Ashitaka E, Hori Y, Sakuma S, Sugimoto T, Sakimura K, Mishina M, Ito S (1999) Involvement of primary afferent C-fibres in touch-evoked pain (allodynia) induced by prostaglandin E2. Eur J Neurosci 11:1849–1856

Mitrovic I, Margeta-Mitrovic M, Bader S, Stoffel M, Jan LY, Basbaum AI (2003) Contribution of GIRK2-mediated postsynaptic signaling to opiate and alpha 2-adrenergic analgesia and analgesic sex differences. Proc Natl Acad Sci USA 100:271–276

Mollereau C, Parmentier M, Mailleux P, Butour JL, Moisand C, Chalon P, Caput D, Vassart G, Meunier JC (1994) ORL1, a novel member of the opioid receptor family. Cloning, functional expression and localization. FEBS Lett 341:33–38

Moore KA, Baba H, Woolf CJ (2002a) Gabapentin-actions on adult superficial dorsal horn neurons. Neuropharmacology 43:1077–1081

Moore KA, Kohno T, Karchewski LA, Scholz J, Baba H, Woolf CJ (2002b) Partial peripheral nerve injury promotes a selective loss of GABAergic inhibition in the superficial dorsal horn of the spinal cord. J Neurosci 22:6724–6731

Morisset V, Urban L (2001) Cannabinoid-induced presynaptic inhibition of glutamatergic EPSCs in substantia gelatinosa neurons of the rat spinal cord. J Neurophysiol 86:40–48

Murakami M, Fleischmann B, De Felipe C, Freichel M, Trost C, Ludwig A, Wissenbach U, Schwegler H, Hofmann F, Hescheler J, Flockerzi V, Cavalie A (2002) Pain perception in mice lacking the β3 subunit of voltage-activated calcium channels. J Biol Chem 277:40342–44051

Muth-Selbach U, Dybek E, Kollosche K, Stegmann JU, Holthusen H, Lipfert P, Zeilhofer HU (2004) The spinal antinociceptive effect of nocistatin in neuropathic rats is blocked by D-serine. Anesthesiology 101:753–758

Nabekura J, Xu TL, Rhee JS, Li JS, Akaike N (1999) Alpha2-adrenoceptor-mediated enhancement of glycine response in rat sacral dorsal commissural neurons. Neuroscience 9:29–41

Nagy GG, Al-Ayyan M, Andrew D, Fukaya M, Watanabe M, Todd AJ (2004) Widespread expression of the AMPA receptor GluR2 subunit at glutamatergic synapses in the rat spinal cord and phosphorylation of GluR1 in response to noxious stimulation revealed with an antigen-unmasking method. J Neurosci 24:5766–5777

Narikawa K, Furue H, Kumamoto E, Yoshimura M (2000) In vivo patch-clamp analysis of IPSCs evoked in rat substantia gelatinosa neurons by cutaneous mechanical stimulation. J Neurophysiol 84:2171–2174

Narumiya S, Sugimoto Y, Ushikubi F (1999) Prostanoid receptors: structures, properties, and functions. Physiol Rev 79:1193–1226

Nebe J, Vanegas H, Neugebauer V, Schaible HG (1997) ω-agatoxin IVA, a P-type calcium channel antagonist, reduces nociceptive processing in spinal cord neurons with input from the inflamed but not from the normal knee joint—an electrophysiological study in the rat in vivo. Eur J Neurosci 9:2193–2201

Nichols ML, Allen BJ, Rogers SD, Ghilardi JR, Honore P, Luger NM, Finke MP, Li J, Lappi DA, Simone DA, Mantyh PW (1999) Transmission of chronic nociception by spinal neurons expressing the substance P receptor. Science 286:1558–1561

Okuda-Ashitaka E, Minami T, Tachibana S, Yoshihara Y, Nishiuchi Y, Kimura T, Ito S (1998) Nocistatin, a peptide that blocks nociceptin action in pain transmission. Nature 392:286–289

Pan YZ, Li DP, Pan HL (2002) Inhibition of glutamatergic synaptic input to spinal lamina II_O neurons by presynaptic α_2-adrenergic receptors. J Neurophysiol 87:1938–1947

Patel S, Naeem S, Kesingland A, Froestl W, Capogna M, Urban L, Fox A (2001) The effects of GABA(B) agonists and gabapentin on mechanical hyperalgesia in models of neuropathic and inflammatory pain in the rat. Pain 90:217–226

Polgar E, Hughes DI, Riddell JS, Maxwell DJ, Puskar Z, Todd AJ (2003) Selective loss of spinal GABAergic or glycinergic neurons is not necessary for development of thermal hyperalgesia in the chronic constriction injury model of neuropathic pain. Pain 104:229–239

Polgar E, Gray S, Riddell JS, Todd AJ (2004) Lack of evidence for significant neuronal loss in laminae I-III of the spinal dorsal horn of the rat in the chronic constriction injury model. Pain 111:144–150

Popratiloff A, Weinberg RJ, Rustioni A (1996) AMPA receptor subunits underlying terminals of fine-caliber primary afferent fibers. J Neurosci 16:3363–3372

Qian J, Brown SD, Carlton SM (1996) Systemic ketamine attenuates nociceptive behaviors in a rat model of peripheral neuropathy. Brain Res 715:51–62

Randic M, Jiang MC, Cerne R (1993) Long-term potentiation and long-term depression of primary afferent neurotransmission in the rat spinal cord. J Neurosci 13:5228–5241

Rees H, Sluka KA, Westlund KN, Willis WD (1995) The role of glutamate and GABA receptors in the generation of dorsal root reflexes by acute arthritis in the anaesthetized rat. J Physiol 484:437–445

Reinold H, Ahmadi S, Depner UB, Layh B, Heindl C, Hamza M, Pahl A, Brune K, Narumiya S, Müller U, Zeilhofer HU (2005) Spinal inflammatory hyperalgesia is mediated by prostaglandin E receptors of the EP2 subtype. J Clin Invest 115:673–679

Reinscheid RK, Nothacker HP, Bourson A, Ardati A, Henningsen RA, Bunzow JR, Grandy DK, Langen H, Monsma FJ Jr, Civelli O (1995) Orphanin FQ: a neuropeptide that activates an opioidlike G protein-coupled receptor. Science 270:792–794

Rhee JS, Wang ZM, Nabekura J, Inoue K, Akaike N (2000) ATP facilitates spontaneous glycinergic IPSC frequency at dissociated rat dorsal horn interneuron synapses. J Physiol 524:471–483

Roche KW, O'Brien RJ, Mammen AL, Bernhardt J, Huganir RL (1996) Characterization of multiple phosphorylation sites on the AMPA receptor GluR1 subunit. Neuron 16:1179–1188

Rudolph U, Möhler H (2004) Analysis of $GABA_A$ receptor function and dissection of the pharmacology of benzodiazepines and general anesthetics through mouse genetics. Annu Rev Pharmacol Toxicol 44:475–498

Ruscheweyh R, Sandkühler J (2002) Role of kainate receptors in nociception. Brain Res Brain Res Rev 40:215–222

Saegusa H, Kurihara T, Zong S, Minowa O, Kazuno A, Han W, Matsuda Y, Yamanaka H, Osanai M, Noda T, Tanabe T (2000) Altered pain responses in mice lacking alpha 1E subunit of the voltage-dependent Ca2+ channel. Proc Natl Acad Sci USA 97:6132–6137

Saegusa H, Kurihara T, Zong S, Kazuno A, Matsuda Y, Nonaka T, Han W, Toriyama H, Tanabe T (2001) Suppression of inflammatory and neuropathic pain symptoms in mice lacking the N-type Ca^{2+} channel. EMBO J 20:2349–2356

Salio C, Fischer J, Franzoni MF, Mackie K, Kaneko T, Conrath M (2001) CB1-cannabinoid and mu-opioid receptor co-localization on postsynaptic target in the rat dorsal horn. Neuroreport 12:3689–3692

Samad TA, Moore KA, Sapirstein A, Billet S, Allchorne A, Poole S, Bonventre JV, Woolf CJ (2001) Interleukin-1beta-mediated induction of Cox-2 in the CNS contributes to inflammatory pain hypersensitivity. Nature 410:471–475

Sandkühler J (2000) Learning and memory in pain pathways. Pain 88:113–118

Sandkühler J, Liu X (1998) Induction of long-term potentiation at spinal synapses by noxious stimulation or nerve injury. Eur J Neurosci 10:2476–2480

Sandkühler J, Chen JG, Cheng G, Randic M (1997) Low-frequency stimulation of afferent Aδ-fibers induces long-term depression at primary afferent synapses with substantia gelatinosa neurons in the rat. J Neurosci 17:6483–6491

Schneider SP, Eckert WA III, Light AR (1998) Opioid-activated postsynaptic, inward rectifying potassium currents in whole cell recordings in substantia gelatinosa neurons. J Neurophysiol 80:2954–2962

Schroeder JE, McCleskey EW (1993) Inhibition of Ca^{2+} currents by a μ-opioid in a defined subset of rat sensory neurons. J Neurosci 13:867–873

Schuler V, Luscher C, Blanchet C, Klix N, Sansig G, Klebs K, Schmutz M, Heid J, Gentry C, Urban L, Fox A, Spooren W, Jaton AL, Vigouret J, Pozza M, Kelly PH, Mosbacher J, Froestl W, Kaslin E, Korn R, Bischoff S, Kaupmann K, van der Putten H, Bettler B (2001) Epilepsy, hyperalgesia, impaired memory, and loss of pre- and postsynaptic $GABA_B$ responses in mice lacking $GABA_{B1}$. Neuron 31:47–58

Schulte G, Robertson B, Fredholm BB, DeLander GE, Shortland P, Molander C (2003) Distribution of antinociceptive adenosine A1 receptors in the spinal cord dorsal horn, and relationship to primary afferents and neuronal subpopulations. Neuroscience 121:907–916

Seeburg PH, Higuchi M, Sprengel R (1998) RNA editing of brain glutamate receptor channels: mechanism and physiology. Brain Res Brain Res Rev 26:217–229

Seltzer Z, Devor M (1979) Ephaptic transmission in chronically damaged peripheral nerves. Neurology 29:1061–1064

Sheen K, Chung JM (1993) Signs of neuropathic pain depend on signals from injured nerve fibers in a rat model. Brain Res 610:62–68

Shimoyama M, Shimoyama N, Hori Y (2000) Gabapentin affects glutamatergic neurotransmission in the rat dorsal horn. Pain 85:405–414

Sivilotti L, Woolf CJ (1994) The contribution of $GABA_A$ and glycine receptors to central sensitization: disinhibition and touch-evoked allodynia in the spinal cord. J Neurophysiol 72:169–179

Sommer C (2003) Painful neuropathies. Curr Opin Neurol 16:623–628

Sorkin LS, Yaksh TL, Doom CM (1999) Mechanical allodynia in rats is blocked by a Ca^{2+} permeable AMPA receptor antagonist. Neuroreport 10:3523–3526

Stanfa LC, Hampton DW, Dickenson AH (2000) Role of Ca^{2+}-permeable non-NMDA glutamate receptors in spinal nociceptive transmission. Neuroreport 11:3199–3202

Szekely JI, Torok K, Mate G (2002) The role of ionotropic glutamate receptors in nociception with special regard to the AMPA binding sites. Curr Pharm Des 8:887–912

Tachibana M, Wenthold RJ, Morioka H, Petralia RS (1994) Light and electron microscopic immunocytochemical localization of AMPA-selective glutamate receptors in the rat spinal cord. J Comp Neurol 344:431–454

Taddese A, Nah SY, McCleskey EW (1995) Selective opioid inhibition of small nociceptive neurons. Science 270:1366–1369

Takeda D, Nakatsuka T, Papke R, Gu JG (2003) Modulation of inhibitory synaptic activity by a non-$\alpha4\beta2$, non-$\alpha7$ subtype of nicotinic receptors in the substantia gelatinosa of adult rat spinal cord. Pain 101:13–23

Tatsuo MA, Salgado JV, Yokoro CM, Duarte ID, Francischi JN (1999) Midazolam-induced hyperalgesia in rats: modulation via $GABA_A$ receptors at supraspinal level. Eur J Pharmacol 370:9–15

Taylor CP (2004) The biology and pharmacology of calcium channel alpha2-delta proteins. CNS Drug Rev 10:183–188

Tognetto M, Amadesi S, Harrison S, Creminon C, Trevisani M, Carreras M, Matera M, Geppetti P, Bianchi A (2001) Anandamide excites central terminals of dorsal root ganglion neurons via vanilloid receptor-1 activation. J Neurosci 21:1104–1109

Travagli RA, Williams JT (1996) Endogenous monoamines inhibit glutamate transmission in the spinal trigeminal nucleus of the guinea-pig. J Physiol 491:177–185

Vanegas H, Schaible HG (2001) Prostaglandins and cyclooxygenases in the spinal cord. Prog Neurobiol 64:327–363

Vasquez E, Bär KJ, Ebersberger A, Klein B, Vanegas H, Schaible HG (2001) Spinal prostaglandins are involved in the development but not the maintenance of inflammation-induced spinal hyperexcitability. J Neurosci 21:9001–9008

Vaughan CW, Connor M, Jennings EA, Marinelli S, Allen RG, Christie MJ (2001) Actions of nociceptin/orphanin FQ and other prepronociceptin products on rat rostral ventromedial medulla neurons in vitro. J Physiol 534:849–859

Wall PD, Woolf CJ (1984) Muscle but not cutaneous C-afferent input produces prolonged increases in the excitability of the flexion reflex in the rat. J Physiol 356:443–458

Westergren I, Nystrom B, Hamberger A, Nordborg C, Johansson BB (1994) Concentrations of amino acids in extracellular fluid after opening of the blood-brain barrier by intracarotid infusion of protamine sulfate. J Neurochem 62:159–165

Willis WD Jr (1999) Dorsal root potentials and dorsal root reflexes: a double-edged sword. Exp Brain Res 124:395–421

Xu TL, Nabekura J, Akaike N (1996) Protein kinase C-mediated enhancement of glycine response in rat sacral dorsal commissural neurones by serotonin. J Physiol 496:491–501

Xu TL, Pang ZP, Li JS, Akaike N (1998) 5-HT potentiation of the $GABA_A$ response in the rat sacral dorsal commissural neurones. Br J Pharmacol 124:779–787

Yang K, Fujita T, Kumamoto E (2004) Adenosine inhibits GABAergic and glycinergic transmission in adult rat substantia gelatinosa neurons. J Neurophysiol 92:2867–2877

Zamanillo D, Sprengel R, Hvalby O, Jensen V, Burnashev N, Rozov A, Kaiser KM, Köster HJ, Borchardt T, Worley P, Lübke J, Frotscher M, Kelly PH, Sommer B, Andersen P, Seeburg PH, Sakmann B (1999) Importance of AMPA receptors for hippocampal synaptic plasticity but not for spatial learning. Science 284:1805–1811

Zeilhofer HU (2005) The glycinergic control of nociception. Cell Mol Life Sci (in press; Jun 17 Epub ahead of print)

Zeilhofer HU, Calò G (2003) Nociceptin/orphanin FQ and its receptor-potential targets for pain therapy? J Pharmacol Exp Ther 306:423–429

Zeilhofer HU, Muth-Selbach U, Gühring H, Erb K, Ahmadi S (2000) Selective suppression of inhibitory synaptic transmission by nocistatin in the rat spinal cord dorsal horn. J Neurosci 20:4922–4999

Zou X, Lin Q, Willis WD (2002) Role of protein kinase A in phosphorylation of NMDA receptor 1 subunits in dorsal horn and spinothalamic tract neurons after intradermal injection of capsaicin in rats. Neuroscience 115:775–786

Rev Physiol Biochem Pharmacol (2005) 155:101–121
DOI 10.1007/s10254-005-0004-5

C. K. Chen

The vertebrate phototransduction cascade: amplification and termination mechanisms

Published online: 11 November 2005
© Springer-Verlag 2005

Abstract The biochemical cascade which transduces light into a neuronal signal in retinal photoreceptors is a heterotrimeric GTP-binding protein (G protein) signaling pathway called phototransduction. Works from psychophysicists, electrophysiologists, biochemists, and geneticists over several decades have come together to shape our understanding of how photon absorption leads to photoreceptor membrane hyperpolarization. The insights of phototransduction provide the foundation for a mechanistic account of signaling from many other G protein-coupled receptors (GPCR) found throughout nature. The application of reverse genetic techniques has strengthened many historic findings and helped to describe this pathway at greater molecular details. However, many important questions remain to be answered.

Abbreviations *AFM:* Atomic force microscopy · *Arr$^{-/-}$:* Homozygous arrestin knockout · *BAPTA:* 1,2-bis(*o*-aminophenoxy)ethane-*N*,*N*,*N'*,*N'*-tetraacetic acid · *cGMP:* 3',5' Cyclic guanosine monophosphate · *CNBD:* Cyclic nucleotide binding domain · *CNG:* Cyclic nucleotide-gated · *DEP:* Disheveled, EGL-10, Pleckstrin · *ERG:* Electroretinography · *EM:* Electron microscopy · *GDP:* Guanosine diphosphate · *GMP:* Guanosine monophosphate · *GTP:* Guanosine triphosphate · *GPCR:* G protein-coupled receptor · *GC:* Guanylate cyclase · *GCAP:* Guanylate cyclase-activating protein · *GCAP$^{-/-}$:* Homozygous GCAP knockout · *GAP:* GTPase-accelerating protein · *GRK:* G protein-coupled receptor kinase · *GGL:* G protein γ-like · *GTPγS:* Guanosine-5'-(γ-thio)-triphosphate · *K$_m$:* Michaelis–Menten constant · *MUV:* Mouse ultraviolet visual pigment · *NCKX:* Na$^+$/Ca^{2+},K$^+$ exchanger · *PDE:* Phosphodiesterase · *P*:* Activated phosphodiesterase · *R:* Rhodopsin · *R*:* Metarhodopsin II (activated form of rhodopsin) · *RCSB:* Research Collaboratory for Structural Bioinformatics · *RK$^{-/-}$:* Homozygous rhodopsin kinase knockout · *Rv$^{-/-}$:* Recoverin knockout · *Rho$^{+/-}$:* Heterozygous rhodopsin knockout · *R9AP:* RGS9 anchoring protein · *RGS:* Regulator of G protein signaling · *RNA:* Ribonucleic acid · *SB:* Schiff base · *SW:* Switch region · *T:* Heterotrimeric transducin · *T*:* Activated

C. K. Chen (✉)
Virginia Commonwealth University, Department of Biochemistry,
1101 E. Marshall Street, Rm 2-032, Richmond, 23298-0614 VA, USA
e-mail: cjchen@vcu.edu · Tel.: 804-828-7973 · Fax: 804-828-1473

transducin · τ_D: Dominant time constant of recovery · τ_{rec}: Recovery constant in dim flash response · V_{max}: Maximal rate of an enzymatic reaction

Introduction

Sensory information about the outside world is sent to the brain in electrical form through streams of action potentials propagating along sensory pathways. In invertebrate vision, light absorption by visual pigments leads to a chain of events resulting in increased membrane conductivity and depolarization (Hardie 2001). In contrast, the coding of light into neural signals in outer segments of vertebrate photoreceptors involves a decrease in membrane conductance and membrane hyperpolarization (Bortoff 1964; Tomita et al. 1967). The light-triggered signal is then relayed at the synapse to downstream neurons by suppressing the tonic release of the neurotransmitter glutamate (Taylor and Smith 2004; Wassle 2004). The presence and performance of photoreceptors is crucial for vision. Patients with photorecep-tor degeneration diseases such as macular degeneration and retinitis pigmentosa suffer from poor vision and eventually total blindness (Rivolta et al. 2002). Individuals with defective phototransduction machinery have to deal with compromised visual functions (Dryja 2000). Because the brain cannot produce visual sensation for photons that are not captured and transduced, the performance of rod and cone photoreceptors sets the ultimate sensitivity and temporal resolution of vision. Rods are quite sensitive to light, but their rate of signal trans-duction is relatively low compared to that of cones. Cones are not as sensitive but are capable of responding to photons carrying different energy and respond faster and in a more adaptive manner (Baylor 1987). Both photoreceptors employ $3',5'$ cyclic guanosine monophosphate (cGMP) as their second messenger. Except for a few molecules that are shared, rods and cones possess their own distinct (but related) sets of gene products to transduce light. This chapter deals mainly with the amplification and termination aspects of phototransduction in rods, which are becoming more accessible than the parallel systems in cones in many experi-mental systems and thus are better understood. Wherever applicable, however, extrapolation of how cone phototransduction may operate will be compared to that of rods. Finally, ex-periments exploiting the kinetic properties of rods and cones will be mentioned to provide insights into how these two types of photoreceptors may differ.

Activation mechanism

The ability of human retinal rods to detect single photons was discovered psychophysi-cally in 1942 (Hecht and Pirenne 1942). It was not until more than 30 years later that this amazing ability was unquestionably established by suction recordings from isolated rods (Lamb et al. 1979; Yau et al. 1979a). For rods to transduce the information of single pho-tons, they have to be quiet in the dark to enhance the signal-to-noise ratio while keeping highly efficient amplification machinery ready for the task (Baylor 1996). Amazingly, rods are able to do so consistently, transducing photon after photon with a stereotypical response characterized by small intertrial fluctuations in response amplitude and duration. This low trial-to-trial variation in the elementary response of rods is noted in sharp contrast to other phenomena involving single particles, such as the time required for the decay of a radioac-tive atom (Field and Rieke 2002; Rieke and Baylor 1998). While a mechanistic account for the origin of the stereotyped single-photon response has yet to be uncovered, the underly-

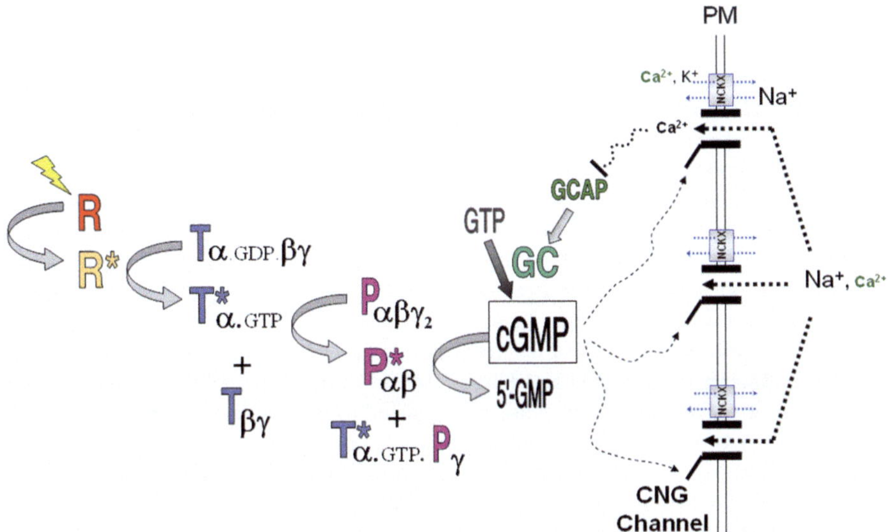

Fig. 1 Phototransduction cascade and the negative Ca^{2+} feedback loop in vertebrate rods. Absorption of photons by rhodopsin (R) leads to the formation of Metarhodopsin II ($R*$), which catalyzes the exchange of GTP for GDP on the α-subunit of heterotrimeric transducin ($T_{\alpha GDP\beta\gamma}$). Activated Tα ($T*_{\alpha GTP}$) dissociates from Tβγ and binds the inhibitory γ-subunit of tetrameric PDE ($P_{\alpha\beta\gamma2}$). The uninhibited catalytic subunit of PDE ($P*_{\alpha\beta}$) hydrolyzes cGMP into 5'-GMP. The drop of intracellular cGMP concentration closes the cyclic nucleotide-gated (CNG) channels located in the plasma membrane (PM), blocks the entry of Na^+ and Ca^{2+}, and results in membrane hyperpolarization. The activation of phototransduction leads to a decline in intracellular Ca^{2+} concentration (the Ca^{2+} signal) because the $Na^+/Ca^{2+},K^+$ exchanger ($NCKX$) is not light sensitive and Ca^{2+} extrusion continues while Ca^{2+} entry through the CNG channel is interrupted by light. The Ca^{2+} signal is sensed by guanylate cyclase-activating protein ($GCAP$), which activates guanylate cyclase (GC) to produce cGMP from GTP. The action of GCAP on GC is sufficient to account for the negative Ca^{2+} feedback on cGMP in the rod's elementary response to light

ing mechanism of signal amplification which turns a molecular event (photon absorption by rhodopsin) into a measurable electrical signal (decrease in membrane conductance) is now pretty well understood. Shown in Fig. 1 is the rod phototransduction mechanism, a typical heterotrimeric G protein signaling pathway which involves a receptor (rhodopsin, or R), a heterotrimeric G protein called transducin, (T), and an effector (cGMP phosphodiesterase, or PDE). Qualitatively speaking, light-activated rhodopsin (R*) turns on transducin by facilitating the exchange of GTP for GDP on its α-subunit (T_α), which in turn activates PDE by interacting with (and thus sequestering) its inhibitory γ-subunit (PDE_γ). Activated PDE hydrolyzes cGMP into 5'-GMP, which leads to a rapid decline in intracellular cGMP concentration, resulting in the closure of the cyclic nucleotide-gated (CNG) channels in the plasma membrane and causing membrane hyperpolarization (Lagnado and Baylor 1992). The relative abundance of R:T:PDE, which are found on the surface of outer segment disc membranes, is approximately 100:10:1 in dark adapted rod outer segments. In the following, we will discuss the amplification events in detail, guided by the molecules involved in each sequential step immediately following photon absorption.

Activation of rhodopsin

The rod visual pigment rhodopsin consists of an apoprotein (opsin) and a covalently attached chromophore (11-*cis*-retinal) at the side chain of Lys296 through a protonated Schiff-base (SB) linkage. In darkness, the 11-*cis*-retinal acts as an inverse agonist which locks rhodopsin into an inactive conformation. The role of light is to isomerize the 11-*cis* bond of the retinal to all-*trans* configuration, a step which occurs within 200 fs (Schoenlein et al. 1991). Photoisomerization of 11-*cis*- to all-*trans*-retinal triggers a series of conformational changes on the protein moiety, at which stage the all-*trans*-retinal behaves like an agonist, inducing rhodopsin to adopt an active conformation called Metarhodopsin II (R*) within a few milliseconds through several spectrally defined intermediates in the sequence of bathorhodopsin (529 nm)→lumirhodopsin (492 nm)→metarhodopsin I (478 nm)↔metarhodopsin II (R*, 380 nm) (Baumann 1976; Lewis and Kliger 1992). As the decay of R* eventually occurs, the all-*trans*-retinal falls off the protein, gets reduced to all-*trans* retinol (Saari et al. 1998), and then gets removed to an adjacent tissue, the retinal pigment epithelium, where it is recycled back to 11-*cis*-retinal through a process called visual cycle (McBee et al. 2001) and is then delivered back to the rod for regeneration of rhodopsin. The covalently bound chromophore and the abundance of visual pigments in the outer segment is a gift from nature that enables studies of the receptor activation mechanism at a level not possible with other G protein-coupled receptors (GPCRs). The positive charge on the protonated SB linkage is energetically unstable inside the protein moiety, but a counterion, Glu113, located on transmembrane helix III stabilizes the base by shifting its pK_a from neutral to alkaline (Sakmar et al. 1989; Sakmar et al. 1991; Steinberg et al. 1993). The presence of the Glu113 counterion is essential for the basic function of rhodopsin, since it prevents the spontaneous hydrolysis of the SB linkage and contributes to the λ_{max} at 498 nm (Nathans 1990; Zhukovsky and Oprian 1989). The activation of rhodopsin also involves a recently identified "counterion switch" mechanism (Birge and Knox 2003) in which Glu181 located in the extracellular loop II (EII) replaces Glu113 to stabilize the SB at the Metarhodopsin I stage, before the eventual deprotonation of the SB linkage, which decreases the pK_a to approximately 6 in R* stage (Ebrey 2000). The three-dimensional structure of rhodopsin in its inactive conformation obtained by X-ray diffraction was shown for the first time in 2000 at high resolution (Palczewski et al. 2000) and so far remains the only available crystal structure (Edwards et al. 2004) of the heptahelical transmembrane GPCR family that constitutes roughly 3% of our genome (Park et al. 2004). The structure can be viewed at the website of the Research Collaboratory for Structural Bioinformatics (RCSB) Protein Data Bank, www.rcsb.org/pdb/ with the PDB ID number 1HZX. In conjunction with studies using a variety of techniques including low-resolution electron density map from cryoelectron microscopy (cryo-EM) (Schertler and Hargrave 1995), site-directed spin labeling (Hubbell et al. 2000), reactivity to thio-specific reagents, and crosslinking (Klein-Seetharaman et al. 1999), the following conformational changes in rhodopsin are inferred to occur immediately following the photoisomerization of 11-*cis*- to all-*trans*-retinal (Hubbell et al. 2003): A cleft at the cytoplasmic end of the helix bundle blossoms up with the separation of transmembrane helices III and VI, accompanied by increased exposure of the inner faces of II, III, VI, and VII, and decreased exposure at the cytoplasmic ends of IV and V. For a more detailed description of rhodopsin activation at the molecular level we will have to wait for the elucidation of the structure of activated rhodopsin. However, there are other things to be done to add to our understanding—e.g., by restricting the rearrangement of the helix bundle through an engineered interhelical crosslink, it is possible to define the minimal helical movement required for activation (Struthers et al. 2000).

Less is known about the chromophore–protein interactions in cone pigments, which also have an acidic residue equivalent to Glu113 of rhodopsin. However, not all the SB linkage of the cone pigments is protonated at ground state (Dukkipati et al. 2002). Ultraviolet-sensitive cone pigments have an unprotonated SB linkage in the dark. In mouse ultraviolet visual pigment (MUV, λ_{max}=357 nm), the unprotonated SB linkage is transiently protonated during the formation of lumirhodopsin and Meta I and reverted to an unprotonated state upon formation of Meta II. Mutagenesis and computational studies suggest that the counterion switch mechanism observed for rhodopsin prior to the formation of Meta II is present during MUV activation (Kusnetzow et al. 2004).

Spontaneous activation of visual pigments in the dark occurs in rods. This dark activity of rhodopsin constitutes the discrete dark noise recorded from toad rods with an average rate of 0.031 s^{-1} at 22°C (Yau et al. 1979b). This frequency of occurrence has been used to derive the half-life of a rhodopsin molecule to be in the order of 1,000 years by assuming that 2×10^9 molecules reside in a toad rod. Temperature dependence of the rate of these dark events reveals an Arrhenius activation energy (Ea) of 23.2±3.2 kcal mol^{-1}. This energy is less than 36 kcal mol^{-1} (Cooper 1979), the energy stored in an early intermediate of light-activated rhodopsin, bathorhodopsin, suggesting that photoexcitation of rhodopsin proceeds over a much larger energy barrier than that caused by thermal isomerization. A specific hypothesis (Barlow et al. 1993) was put forward to suggest that only chromophore with unprotonated SB linkages to Lys296 will undergo thermal isomerization at an appreciable rate. Because of the high pK_a of the protonated SB linkage, few rhodopsins would be capable of undergoing isomerization in any brief interval. The hypothesis is based upon theoretical calculations as well as experimental observations on the pH dependence of the thermal event rate in Limulus photoreceptors. The behavior of the E113Q rhodopsin mutant is in agreement with the hypothesis. In this mutant, the counterion Glu113 is changed to a neutral glutamine, reducing the pK_a of the SB nitrogen to about 7 (Lin et al. 1992; Sakmar et al. 1991). This mutation causes E113Q rhodopsin to activate the G protein without light exposure (Robinson et al. 1992). This hypothesis was further tested in the salamander L cone, whose pigment has a comparable Ea for thermo-isomerization to that of rods, but the pK_a of the SB linkage is much lower (Liang et al. 1994). It was found that manipulations designed to affect the protonation status of the SB linkage have no effect on the rate of occurrence of the elementary response in L cone (Sampath and Baylor 2002). An alternative explanation was put forward recently (Ala-Laurila et al. 2004), taking into account the internal energy present in many vibrational modes of the rhodopsin molecule, that thermoactivation and photoactivation energy barriers of visual pigments are quite similar to each other and that activation of rhodopsin by heat or by light may follow exactly the same molecular route.

It is well known in macaque retina that cones have more dark noise than rods (Baylor et al. 1984; Schnapf et al. 1990). Why are cones noisier? The answer may be the pigment. The implication that the sensitivity differences between rods and cones are attributable to the pigments themselves was elegantly demonstrated by expressing human and salamander long-wavelength (red) sensitive cone pigments in *Xenopus* rods, and human rod pigment in *Xenopus* cones (Kefalov et al. 2003). These genetic manipulations showed that altering the pigment alone was sufficient to alter the noise level and not the kinetics of phototransduction. Expressing a red cone pigment in rod converts a relatively quiet rod into a noisy one. The rate of spontaneous photoisomerization-like events in cone pigments was calculated to be over 10,000 times that of rod pigments. The molecular basis of this dramatic difference is unclear. It was suggested that the chromophore-binding pocket in cone pigment is much looser than that of the rhodopsin, allowing rapid conformational changes and recovery from bleaching which requires the dissociation of the protein moiety and the chromophore followed by filling of the pocket with re-isomerized 11-*cis*-retinal (Osorio and Nilsson 2004). Cone pigment has another distinguishable property unseen in rhodopsin in which the SB linkage may be spontaneously broken without isomerization (Crescitelli 1984; Matsumoto et al. 1975). This reversible covalent association between cone opsin and 11-*cis*-retinal is apparently a general property of all cone pigments in salamander and, when studied in detail in red cones, it was surprisingly found that about 10% of the pigments in dark adapted state are empty at the chromophore binding pocket. This large fraction of free opsin and its constitutive activity toward transducin leads to an apparent twofold desensitization in red cones, on top of the desensitization caused by thermo-isomerization of red cone pigment. Together, the unique properties of cone pigment make the dark adapted cone behave to some extent like a light-adapted rod, manifested as reduced sensitivity and accelerated response kinetics (Kefalov et al. 2005; Travis 2005).

A provocative presentation of images of rows of rhodopsin dimers in rod disc membranes by atomic force microscopy (AFM) (Fotiadis et al. 2003), regarded by many as a milestone in GPCR research (Bulenger et al. 2005), has surprisingly received less applause from phototransduction scientists who have long considered non-activated rhodopsin as a monomer (Chabre et al. 2003). The major reservation seems to be directed more against the ordered paracrystalline array presented, rather than against the notion of rhodopsin existing as a dimer. Photodichroism studies revealed that rhodopsin undergoes rapid rotational diffusion in situ (Cone 1972). The kinetics of flash-bleaching recovery indicates that it also has fairly rapid lateral diffusion (Poo and Cone 1974). The image of a crowded and rigid packing of rhodopsin in the paracrystalline array on the disc surface is apparently at odds with these and many other previous findings. The notion of a non-activated dimeric (or higher oligomeric) rhodopsin on the other hand has been experimentally supported in recent reports (Jastrzebska et al. 2004; Suda et al. 2004). The challenge now is to determine whether or not rhodopsin is a monomer in vivo under conditions where electrophysiological data are collected. If proved, this will invoke the need to modify our current description of the vertebrate phototransduction cascade. If not, despite mounting evidence of GPCR oligomerization and its roles in signaling, biosynthesis, maturation, and degradation (Bulenger et al. 2005), monomeric rhodopsin may represent an exception. This is not unprecedented in vertebrate phototransduction where surprises, such as membrane hyperpolarization and a decrease in intracellular Ca^{2+} concentration upon receptor activation, have been documented.

Activation of transducin and phosphodiesterase

The C-terminal tail as well as the second and third intracellular loops of rhodopsin have been shown by many studies to be involved in the activation of transducin (Preininger and Hamm 2004). The exchange of GTP for GDP on the α-subunit of transducin, enzymatically catalyzed by R*, serves as the first amplification step in the phototransduction cascade. To estimate the gain factor, the kinetic parameters of the metarhodopsin II–transducin interaction have been analyzed by several techniques including an indirect method measuring flash-induced change in infrared light-scattering (Vuong et al. 1984). Transducin activation measured this way has a fairly fast guanine nucleotide exchange rate at greater than 1,000 S^{-1}. Therefore, an unquenched R* during its effective lifetime can activate hundreds of transducin molecules in rods. However, direct biochemical measurements of the exchange rate, e.g., by measuring the binding of the amount of nonhydrolyzable GTP analog to transducin immobilized onto nitrocellulose filters (Fung et al. 1981), always yield smaller values. The difference has been about tenfold between direct and indirect measurements in the literature (Pugh and Lamb 1993). Further biochemical experiments addressing the paradoxical findings were recently attempted (Leskov et al. 2000), but the issue remains unresolved.

Mutagenesis studies have identified residues in the $\alpha4/\beta6$ loop, especially residues at the C-terminal end of the $\alpha4$ helix, to play an important role in receptor-mediated activation of transducin (Natochin et al. 1999). The extreme C-terminus of transducin was also implicated as being involved in receptor interaction (Aris et al. 2001; Natochin et al. 2000). Crosslinking studies demonstrated that residue 240 in the third intracellular loop of rhodopsin crosslinks to a region on Tα encompassing residues 342–345 upon rhodopsin activation (Cai et al. 2001). The exact structural transitions linking the changes induced by photon absorption on rhodopsin to the changes in the nucleotide-binding pocket of Tα remain unclear. Conceivably, the structure of an activated rhodopsin and transducin complex will provide much-needed details. In contrast to the lack of structural details underlying transducin activation, the conformational changes of transducin after activation are better understood (Preininger and Hamm 2004; Sprang 1997). Most of the information is derived from comparisons of the crystal structures of the transducin α-subunit in its heterotrimeric inactivated (Lambright et al. 1996) and monomeric activated forms (Noel et al. 1993) as well as in a state mimicking transitional GTP hydrolysis (Sondek et al. 1994). The GTPase domain of transducin is similar to those of small G proteins such as Ras. The guanine nucleotide binds in a cleft between the GTPase domain and an additional helical domain. The major structural feature of the interaction between Tα and GDP is based on interactions of conserved residues between the $\beta1$ strand and $\alpha1$ helix in the GTPase domain and the α- and β-phosphates of GDP. A salt bridge of R174/E39 augments the binding of transducin to GDP (Preininger and Hamm 2004). The ground state transducin structure is very stable and, as a result, spontaneous GDP/GTP exchange occurs with a measured rate of 0.0001 S^{-1} (Ramdas et al. 1991). Upon activation by R*, transducin loses the bound GDP and becomes temporarily empty at its nucleotide-binding pocket. Upon subsequent binding of GTP, three parts of the molecule called the "switch region" (SW I, SW II, SW III) change conformation and can be readily identified when comparing inactivated and activated structures. GTP-bound Tα has Mg^{2+} coordinated to oxygens of the β- and γ-phosphates of the bound nucleotide. In a structure with a bound nonhydrolyzable GTPγS, the activated state of transducin is stabilized by interactions between basic residues of the SW II region with acidic residues at the SW III region, along with interactions between the SW II region and the $\alpha3$ helix. In this activated state, Tα dissociates from the T$\beta\gamma$ dimer and from the R*. It then forms a tight complex with the γ-subunit of PDE.

Rod PDE belongs to the PDE6 family and is a tetrameric membrane-associated protein (Cote 2004). The holoenzyme consists of two equally active and highly similar catalytic subunits α and β, and two identical inhibitory γ subunits (Baehr et al. 1979; Hurley and Stryer 1982). Each catalytic subunit is posttranslationally modified at its C-terminus with an isoprenyl group that is responsible for its high affinity association with the outer segment disc membrane (Li et al. 1990; Qin et al. 1992). Each catalytic subunit has a catalytic domain responsible for cGMP hydrolysis and two regulatory GAF domains, one of which binds cGMP with high affinity (Mou and Cote 2001). The roles of these noncatalytic cGMP binding sites in PDE activity or in phototransduction remain poorly defined. Occupancy of these sites with cGMP was reported to increase the binding affinity of PDEγ for PDE$\alpha\beta$ (Yamazaki 2002). The interaction between Tα in its GTP bound form with PDEγ relieves its inhibitory constraint on the two catalytic subunits. It takes two activated Tα to fully activate a holo PDE in vitro. The activation of PDE by transducin constitutes no gain during phototransduction. It is the catalytic activity of PDE that provides the second gain step. It was reported that activated PDE operates with nearly perfect efficiency, with a newly determined K_m of approximately 10 µM and a k_{cat} of 2,200 S^{-1} (Leskov et al. 2000). This gives a catalytic power (k_{cat}/K_m) of $2{\times}10^8$ M^{-1}S^{-1} per subunit, 30-fold more powerful at hydrolyzing cGMP than previously anticipated (Stryer 1991).

cGMP-gated cation channel

In rod and cone photoreceptors, the decrease in concentration of cGMP leads to the closure of CNG channels located in the plasma membrane of the outer segment (Cook et al. 1989; Kaupp et al. 1988), which hyperpolarizes the entire cell. The hyperpolarization relays visual information generated at the outer segment to the synaptic terminal, where it decreases the tonic release of the neurotransmitter glutamate. These channels are part of a subfamily of ion channels that are regulated by the direct binding of cyclic nucleotides. Like their voltage-activated counterparts, CNG channels have six membrane-spanning domains and internal N- and C-terminal regions; however, they are only weakly activated by voltage (Kaupp et al. 1989). Instead, CNG channels are gated by the direct binding of cGMP (Fesenko et al. 1985) through a C-terminal cyclic nucleotide-binding domain (CNBD). The value of $K_{1/2}$, the concentration of cGMP necessary to half-maximally activate the rod CNG channel, is 10–80 µM (Fesenko et al. 1985; Nakatani and Yau 1988; Yau and Nakatani 1985b). This value is higher than the free intracellular cGMP concentration (3–10 µM) in the dark (Nakatani and Yau 1988; Pugh and Lamb 1993). The relatively low affinity of rod CNG channels for cGMP leads to a fast off-rate for ligand binding and allows the channel to close within a few milliseconds in response to a light-induced drop of cGMP concentration. Native rod CNG channels comprise CNGA1 subunits and CNGB1 subunits. In cones, the channels consist of CNGA3 and CNGB3 subunits (Johnson et al. 2004). These channel subunits are 35% identical on the amino acid level but are functionally different. CNGA1 subunits form homomeric CNG channels in heterologous systems with some of the properties of native rod channels (Kaupp et al. 1989). CNGB1 subunits do not form functional homomeric channels; however, coexpression of CNGA1 and CNGB1 subunits results in heteromeric channels that recapitulate many of the biophysical and pharmacological properties of native rod channels (Chen et al. 1993; Korschen et al. 1995). The CNGB1 subunit confers several new properties on the heteromeric channels, increased permeability of Ca^{2+} relative to Na^+, flickery gating kinetics, sensitivity to the blocker l-*cis*-diltiazem, and sensitivity to inhibition by Ca^{2+}-calmodulin. Surprisingly, these properties are produced by only a single CNGB1 subunit, as rod channels contain three CNGA1 subunits and only one CNGB1 subunit (Zimmerman 2002; Weitz et al. 2002; Zheng et al. 2002; Zhong et al. 2002).

The light-insensitive and electrogenic Na^+/Ca^+-K^+ exchanger (NCKX) was reported to be in close enough proximity to be crosslinked by thio-specific reagents in native rod outer segment membranes (Schwarzer et al. 2000). This implies that NCKX may form a stable complex with CNGA1. This notion is supported by coimmunoprecipitation of heterologously expressed NCKX and CNG channels (Kang et al. 2003). Several physiological implications of this finding are noteworthy. First, this may be a mechanism required to regulate protein synthesis and establish a fixed molar ratio of both CNG channels and NCKX in the cell. Because in the outer segment the CNG channel is the only influx route of Ca^{2+} and NCKX the only efflux, the ratio of the two critically controls the free Ca^{2+}-concentration in photoreceptors in the dark, and the rate of Ca^{2+} declines upon illumination. Second, the light-induced drop in intracellular Ca^{2+} (the Ca^{2+} signal, see below, "The feedback between Ca^{2+} and cGMP") may be restricted to the immediate vicinity of the NCKX/CNG channel complex. Further investigations into the structure/function aspects of this interaction and its putative roles in phototransduction are warranted and shall provide valuable insights.

The rate-limiting step of activation

Several steps involved in amplification are known to have very fast kinetics. The transition from photon absorption to the generation of R* occurs in a few milliseconds. The fast response time (also a few milliseconds) of the CNG channels on excised outer segment patches to cGMP and the newly determined catalytic power of light-activated PDE$\alpha\beta$ indicate that the rise of the single photon response is rate-limited somewhere between the activation of transducin by R* and subsequent activation of PDE. The rate of transducin activation depends on the rate of its binding to R* as well as on the rate of exchange of GTP for GDP. The indirect light-scattering measurement of this process suggested that guanine nucleotide exchange is a slower step than the diffusional encounter of R* and transducin (Vuong et al. 1984). This is not in accordance with a recent finding from recordings of mouse rods expressing half the normal level of rhodopsin (Rho$^{+/-}$ rods). Photoreceptors of Rho$^{+/-}$ animals appeared normal and healthy and photoresponses from these cells rose at twice the normal rate. It was argued that a lower rhodopsin concentration reduces protein crowding on the disc membrane, thereby increases rhodopsin's diffusion coefficient and its rate of encounter of transducin (Calvert et al. 2001). Adding to the paradox was a re-examination of the Rho$^{+/-}$ rods by Liang et al. who reported an approximate 40% reduction in rod outer segment volume, rhodopsin content, as well as 11-*cis*-retinal level in these cells (Liang et al. 2004). The AFM images of Rho$^{+/-}$ rod disc prepared and imaged at room temperature showed a large area of lipid, while rhodopsin existed in small raft-like structures as well as in large and organized paracrystalline. The authors suggested that the observed acceleration of phototransduction in Rho$^{+/-}$ rods was not due to a lower density of rhodopsin on the disc surface but to the structural changes of the whole rod outer segment (ROS). Thus, the identity of the rate-limiting step of activation remains undefined.

Deactivation mechanism

The recovery of dark current in single photon responses occurs reproducibly within a few hundred milliseconds in mammalian rods. Unlike the sequential events leading from photon absorption by rhodopsin to rapid decline in cGMP levels during activation, deactivation requires rapid restoration of cGMP and the concerted quenching of active intermediates, namely R*, T*, and P*, produced during photoexcitation. Significant progress has been made in the past decade toward our understanding of the timing and the molecules crucial for these deactivation steps, which in essence are important for good temporal resolution and reliable photon counting. Recent experimental approaches in the identification of the rate-limiting step in rod recovery may pave the way to further elucidate the molecular basis of rod/cone differences in their sensitivity and response kinetics to light.

The feedback between Ca^{2+} and cGMP

It was once hypothesized that light-released Ca^{2+} from discs and that increased intracellular calcium then blocked membrane channels to decrease surface conductance (Hagins and Yoshikami 1974). This so-called "calcium hypothesis" lost its steam around 1985 when evidence supporting the rival "cGMP cascade theory" started to emerge. Fesenko et al. showed that the conductivity of an isolated patch of outer segment membrane can be rapidly and reversibly raised by physiological levels of cGMP and not by increased Ca^{2+} (Fesenko et al. 1985). It was later shown by Yau and Nakatani while examining the exchange current seen on the plateau of a flash response that light actually decreases rather than increases the intracellular Ca^{2+} (Yau and Nakatani 1985a). During the course of a flash response, the influx of Ca^{2+} into outer segment declines as a result of channel closure. This is because the exchange current is not light sensitive and continues to utilize the sodium gradient to extrude intracellular Ca^{2+} during channel closure. The light-induced decline of intracellular Ca^{2+} (the so-called Ca^{2+} signal) has been measured by several laboratories using different techniques. It drops from approximately 250–700 nM in the dark to less than 50 nM upon illumination (Gray-Keller and Detwiler 1994; Sampath et al. 1999; Woodruff et al. 2002). The Ca^{2+} signal triggers coordinated changes that negatively regulate the cell's response to light. It can decrease rhodopsin activity by allowing light-dependent rhodopsin phosphorylation (Chen et al. 1995a; Kawamura 1993; Klenchin et al. 1995) and increase the sensitivity of the channel for cGMP by relieving calmodulin restraint on channel activity (Hsu and Molday 1993). In addition, it can increase the activity of guanylate cyclases (GCs) through the action of the guanylate cyclase-activating proteins (GCAPs) (Dizhoor et al. 1995; Palczewski et al. 1994). Due to its multifaceted actions on many reactions underlying phototransduction, it is not surprising to find that it is centrally involved in photoreceptor light adaptation (Koutalos et al. 1995a, b), a topic that is not a focus of this chapter. The readers, however, are encouraged to follow several publications (Burns and Baylor 2001; Calvert and Makino 2002; Nakatani et al. 2002) for a deeper and broader appreciation of this aspect.

What is the relative contribution of the above-mentioned reactions in the negative feedback loop between Ca^{2+} and cGMP in photoreceptors? In 1988, a highly cooperative control of cGMP synthesis by calcium in rod outer segment extract was reported (Koch and Stryer 1988). This led to the identification of recoverin (Dizhoor et al. 1991; Lambrecht and Koch 1991) and GCAPs (Dizhoor et al. 1995; Palczewski et al. 1994) in subsequent years as potential mediators of this unique effect in rod outer segment extract. While the candidacy of recoverin was removed (Hurley and Chen 2001; Hurley et al. 1993), GCAPs on the other hand withstood the test of time (Palczewski et al. 2004). To study the function of GC regulation by GCAPs, mice lacking functional GCAPs were generated to enable genetic examination of the negative feedback loop between Ca^{2+} and cGMP (Mendez and Chen 2002). By examining the effect of the Ca^{2+} buffer BAPTA [1,2-bis(*o*-aminophenoxy)ethane-*N*,*N*,*N'*,*N'*-tetraacetic acid] in the GCAP$^{-/-}$ and control rods, it was found that during the course of elementary light response, regulation of GC is the only operational Ca^{2+} feedback loop in these cells (Burns et al. 2002). By analyses of the dark noise and the light response of GCAP$^{-/-}$ rods and careful comparison with those from normal rod, Burns et al. concluded that GC responds to light-induced Ca^{2+} decline as early as 40 ms after photoexcitation of rhodopsin and in a highly cooperative manner (Hill coefficient of ~4). The high cooperativity suggests that GC and/or GCAP may function in a dimeric state in rods. The authors also noted a comparable rate of recovery during saturated responses between GCAP$^{-/-}$ and control rods. This echoes the finding (see below) that the negative feedback loop between Ca^{2+} and cGMP does not rate-limit the recovery of rod's response to light.

What is then the role of recoverin? Recoverin inhibits light-dependent rhodopsin phosphorylation in a Ca^{2+}-dependent manner by binding to rhodopsin kinase (RK) in vitro (Chen 2002). Consistent with such an effect, internal dialysis of exogenous Ca^{2+}-recoverin into intact rod outer segments delayed the recovery of the flash response (Gray-Keller et al. 1993). Addition of exogenous recoverin was reported to lengthen the response duration by prolonging the lifetime of catalytically active, photoexcited rhodopsin (Erickson et al. 1998). These results were consistent with the notion that recoverin imparts Ca^{2+} dependence to the shutoff of rhodopsin by RK. However, experiments on isolated outer segments permeabilized with staphylococcal α-toxin produced no support that the level of Ca^{2+} controlled the light-dependent phosphorylation of rhodopsin (Otto-Bruc et al. 1998). To settle these contradictory findings, recoverin knockout ($Rv^{-/-}$) rods were recorded and analyzed to indicate that Ca^{2+}-recoverin prolongs the dark-adapted flash response and increases the rods' sensitivity to dim steady light. It is also found that $Rv^{-/-}$ rods have faster Ca^{2+} dynamics, indicating that recoverin can act as a Ca^{2+} buffer in the outer segment (Makino et al. 2004). These data support the notion that Ca^{2+}-recoverin may potentiate light-triggered phosphodiesterase activity, probably by effectively prolonging the catalytic activity of activated rhodopsin. Recoverin is present in other retinal cells and has recently been reported to enhance signal transmission in mouse retina (Sampath et al. 2005).

Rhodopsin deactivation

The half life of R* measured in vitro is too long to account for the rapid recovery of rod photoresponse. Reconstituted biochemical studies have suggested a two-step process for timely quenching of R*, namely phosphorylation by RK followed by the binding of arrestin (Wilden et al. 1986). The requirement of phosphorylation for effective R* deactivation has been demonstrated in vivo by targeted deletion of RK (Chen et al. 1999; Lyubarsky et al. 2000) or by removing/substituting the phosphorylation sites located at rhodopsin's C-terminal tail (Chen et al. 1995b; Mendez et al. 2000). It was shown that reproducible quenching of R* requires at least three phosphorylation events and that all six sites need to be phosphorylated to ensure normal recovery. In transgenic frog rods expressing a human red cone opsin in which all C-terminal phosphorylation sites were mutated, a prolonged photoresponse was observed (Kefalov et al. 2003). This suggests that the two-step quenching mechanism can be applied to cone pigment, despite the fact that the half life of Meta II is more than 50 times shorter for cone opsin than for R* (Imai et al. 1997; Shichida et al. 1994), and despite the suggestion that cone pigment regeneration, rather than phosphorylation, mediates cone recovery (Cideciyan et al. 1998).

Following the phosphorylation of R*, the protein arrestin binds and quenches its remaining catalytic activity. Mouse rods lacking arrestin ($Arr^{-/-}$) display an abnormal photoresponse with a long recovery phase with a time constant of approximately 45 s (Xu et al. 1997). Interestingly, in $RK^{-/-}$ rods where single-photon responses are highly variable and prolonged, the averaged half-life of R*, measured by fitting the recovery phase of dim flash responses with a single exponential function, is approximately 3–4 s (Chen et al. 1999). In rods, a phosphorylation-independent shutoff mechanism exists whose molecular identities have yet to be uncovered. Arrestin gene produces at least two splice variants, the full-length arrestin (p48) and a C-terminal truncated form (p44) (Smith et al. 1994). Although less abounding, p44 has been shown to possess higher affinity for phosphorylated rhodopsin and to quench photoexcited rhodopsin more efficiently in vitro (Langlois et al. 1996). Which form of arrestin mediates rhodopsin deactivation in normal rods is unclear. Cones express their own arrestin called cone arrestin or X-arrestin (Craft and Whitmore 1995). The role of cone arrestin in cone recovery is postulated to resemble the action of its counterpart in rods (Zhu et al. 2003), but this has yet to be demonstrated in vivo.

In humans, mutations in RK and arrestin genes cause Oguchi disease, which is a recessive form of congenital stationary night blindness (Cideciyan et al. 1998; Fuchs et al. 1995; Yamamoto et al. 1997). The resulting defect in R* deactivation manifests itself as night blindness in patients who have not reported problems with their daytime vision. Unlike arrestin, RK is found in rod and cone photoreceptors in humans and mice (Chen et al. 1999; Zhao et al. 1997). Electroretinography (ERG) analyses revealed a cone recovery defect in $RK^{-/-}$ mice but not in $Arr^{-/-}$ mice (Lyubarsky et al. 2000). Why do Oguchi patients with a defective RK gene have normal daytime vision? It is now evident that human cones (and not mouse cones) express an additional closely related kinase called GRK7 (Chen et al. 2001; Weiss et al. 2001). This may be the reason why daytime vision of Oguchi patients with defective RK alleles remains normal, as GRK7 may compensate for defective RK to quench photoexcited cone pigments. It is intriguing that human cones possess two closely related G protein-coupled receptor kinases (GRKs). It should be noted, however, that this expression pattern is not prevalent and exists only in selected species (Weiss et al. 2001). Interestingly, the reported kinetic properties between RK and GRK7 remain controversial. One report demonstrates GRK7 to be a superior enzyme (Tachibanaki et al. 2005) while another report shows a similar V_{max} and K_m for these two enzymes (Horner et al. 2005). Since mouse photoreceptors express and rely exclusively on RK to shut off visual pigments, ectopic expression of human GRK7 in mice may thus provide an opportunity to examine in vivo the advantage of having two GRKs in photoreceptors.

Deactivation of transducin and phosphodiesterase

Recovery of photoresponse will not be complete without timely deactivation of transducin and PDE. Like other heterotrimeric G proteins, the intrinsic GTPase activity of Tα hydrolyzes the bound GTP to GDP to turn itself off. Upon GTP hydrolysis, PDEγ is released from Tα, which then re-inhibits PDE's catalytic subunits. The rate at which GTP is hydrolyzed by Tα thus determines the rate at which PDE is deactivated.

Transducin's slow intrinsic GTPase activity assayed in vitro is not compatible with the time course of photoreceptor flash responses (Bourne and Stryer 1992). This prompted a search for a GTPase-accelerating protein (GAP) for transducin in rod outer segment extracts about a decade and a half ago and led to the identification of PDEγ as a potential GAP (Arshavsky and Bownds 1992). The assignment of PDEγ as the GAP for transducin seemed plausible, as it imparts a safety mechanism to prevent transducin from being short-circuited without activating its effector (Bourne and Stryer 1992). An engineered W70A mutant in PDEγ, which impairs its interaction with transducin (Slepak et al. 1995), reduces the rate of transducin GTP hydrolysis and delays the recovery of dim flash responses in transgenic mouse rods (Tsang et al. 1998). However, recombinant PDEγ has no GAP activity (Angleson and Wensel 1993) and it is now known that PDEγ does not posses GAP activity in the absence of RGS9-1 (Chen et al. 2000; He et al. 1998), a member of the regulator of G protein signaling (RGS) family (Ross and Wilkie 2000). RGS9-1 complexes with Gβ5-L through the G protein γ-like (GGL) domain (Makino et al. 1999) and interacts with its membrane anchors R9AP (RGS9-1 anchoring protein) (Hu and Wensel 2002) through the N-terminal DEP (Dishevell/Egl10/Pleckstrin) domain (Martemyanov et al. 2003). One should now view transducin GAP as a protein complex consisting of RGS9-1/Gβ5-L/R9AP, since rods lacking Gβ5-L, R9AP, or RGS9-1 display similar delay in the recovery phase of their flash responses without much noticeable changes during activation (Keresztes et al. 2004; Krispel et al. 2003b). This GAP complex stimulates the intrinsic transducin GTPase only when transducin is bound to PDEγ. The RGS9-1/Gβ5-L/R9AP complex is present in both rod and cone outer segments and is one of the few molecules that is used in both rod and cone phototransduction. The concentration of this GAP complex is much higher in cones than in rods (Cowan et al. 1998; Zhang et al. 1999; Zhang et al. 2003), and this is one reason why cones respond to light with faster response kinetics (see the following section).

Loss-of-function mutations in human RGS9 or R9AP genes lead to a recessive disease called bradyopsia (Nishiguchi et al. 2004). These patients have problems adjusting to sudden changes in ambient light intensity, seeing fast moving objects, and playing ball games, indicating that cone cells rely on the RGS9-1/Gβ5-L/R9AP GAP complex for good temporal resolution at daytime. The recovery of cone-derived electroretinographic responses was greatly delayed in RGS9$^{-/-}$ mice (Lyubarsky et al. 2001). Interestingly, bradyopsia patients do not report problems with their dim-light vision. This suggests that rods may have a backup mechanism to compensate for the loss of the RGS9-1/Gβ5-L/R9AP GAP complex. It was noted in RGS9$^{-/-}$ rods that a weak GAP activity is operational at dim light level or as bright flash response recovers to a point where intracellular Ca^{2+} or cGMP concentration approaches its dark level (Chen et al. 2000). The identity of this weak GAP activity is unclear.

Photoreceptors process the transcripts of RGS9 and Gβ5 genes differently from other neurons of the central nervous system. A splice variant of the RGS9 transcript called RGS9-2 was found in striatal neurons (Rahman et al. 1999; Zhang et al. 1999). Similarly, another splice variant of the Gβ5 transcript called Gβ5-S is found in brain and in inner retina (Watson et al. 1996; Watson et al. 1994). It is not yet clear how this is achieved in photoreceptors or why. RGS9-1 was shown to be phosphorylated in the dark (Balasubramanian et al. 2001; Hu et al. 2001). The physiological role of RGS9-1 phosphorylation has not been explored in vivo. Another interesting feature of the transducin GAP complex is that its constituting components are obligate partners for stability. Inactivation of any one component in mouse results in an increased instability of the other two through a posttranscriptional mechanism, presumably through regulation of protein degradation (Chen et al. 2000, 2003; Keresztes et al. 2004). Such a relationship is not seen in other photoreceptor protein complexes such as the catalytic subunits of the cGMP-phosphodiesterase in mouse rods. In rd1 mice carrying a non-sense mutation in the β-subunit of PDE, the α-subunit is still present in an inactive form (Lee et al. 1988). The structural aspects of this RGS9-1/Gβ5-L/R9AP protein complex are just beginning to be appreciated, while other properties of this complex such as the regulation of complex assembly and its transport to the outer segment have yet to be studied.

Rate-limiting step of recovery in rod phototransduction

Of all the reactions mentioned above required for timely deactivation of photoexcited intermediates and restoration of cGMP, the slowest step will dominate the shape of the recovery phase. This is evident by the fact that the recovery phase of dim flash responses can be fitted by a single exponential function whose time constant (τ_{rec}) reflects the rate of the slowest step (Burns and Baylor 2001; Krispel et al. 2003a). Another way of looking at this issue is to determine the so-called dominant time constant (τ_D) of recovery (Lyubarsky et al. 1996; Pepperberg et al. 1992). This is done by plotting the time the responses stay in saturation vs the natural logarithm of the intensities of the eliciting flashes. The slope of the plot gives the τ_D. In amphibian rods, τ_D is about 2 s, and in mouse rods the value is about 200 ms. The value of τ_D and τ_{rec} is identical in mouse rods when flash intensity is below roughly 4,000 photons/μm^2. Stronger flashes increase the value of τ_D, indicating that as the flash strength exceeds approximately 1,000 photoisomerizations per rod, a new factor starts to dominate the recovery phase. It is not clear whether this occurs in the same step which dominates the recovery to dim light or in a different one. Very importantly, it was determined that τ_D is invariant for a range of flash strengths over which GC activity varies about tenfold (Lyubarsky et al. 1996) and that in GCAP$^{-/-}$ rods the τ_D is similar to that of control rods

(Burns et al. 2002). This eliminates the involvement of the Ca^{2+} negative feedback loop as a candidate for the rate-limiting step.

Experiments aiming to eliminate additional steps as being rate-limiting have not yielded unequivocal results. In truncated salamander rods by substituting nonhydrolyzable GTPγS for GTP, it was reported that the falling phase of the dim flash response is greatly prolonged while the rising phase and the peak amplitude remain virtually unchanged. It was then concluded that rhodopsin activity must have subsided enough by the time response peaks, and that rhodopsin deactivation should not be considered as the rate-limiting step of recovery (Sagoo and Lagnado 1997). In truncated toad rods, however, the direct assessment of rhodopsin's catalytic activity by rapidly changing the GTP concentration inside the recorded cell revealed that R* decayed with a time constant of 2 s, which is identical to τ_D determined in intact toad rods, indicating that R* deactivation may not be eliminated as the rate-limiting step (Rieke and Baylor 1998). In mouse rods, a novel form of adaptation was reported (Krispel et al. 2003a) which acts by speeding the recovery (and not by reducing the gain) of phototransduction. This form of adaptation (sometimes referred to as "Burns adaptation"), induced by a 3-min illumination of just-saturating steady background light, persists with a roughly 80-s decay constant immediately following the termination of adapting background illumination. Interestingly, the Burns adaptation is fully operational in $RGS9^{-/-}$ rods in which the rate-limiting step is known to be transducin turnoff. This suggests that the deactivation of transducin can be accelerated in rods under those recording conditions. This brings up the interesting idea of identifying the rate-limiting step of recovery under normal conditions, which is to identify it not by eliminating what is not but by directly accelerating it. This is feasible by either elevating the levels of proteins responsible for each deactivation step or by substituting the endogenous ones with dominant mutants possessing enhanced enzymatic activities. Transgenic mice with an overexpressed transducin RGS9-1/Gβ5-L/R9AP complex or RK in retinal rods have recently been generated and characterized for this purpose. It has been found that by elevating the concentration of transducin GAP complex to different levels, the recovery of flash responses of rods can be accelerated in a dose-dependent manner. Conversely, an elevation of the RK level in mouse rods produces no such effects on their recovery from light. These findings unequivocally identify transducin deactivation as the rate-limiting step of recovery in rod phototransduction (C.M. Krispel, D. Chen, Y.-J. Chen, K.A. Martemyanov, N. Quillinan, V.Y. Arshavsky, T.G. Wensel, C.-K. Chen, and M.E. Burns, submitted).

The identification of the rate-limiting step during the recovery of rods from light has important implications. First, cones from several species have been shown to express higher levels of this transducin GAP complex (see above, "Deactivation of transducin and phosphodiesterase"), and the higher level of the transducin GAP complex has been postulated to account for the faster recovery of cone photoresponses. This prediction has now been proved correct. Furthermore, the level of the transducin GAP complex in cones is tenfold higher than that in rods (Zhang et al. 2003), suggesting that the deactivation of transducin may already operate at its full speed and that the rate-limiting step in cone recovery may no longer be transducin shutoff, but has shifted to another deactivation step. The second-slowest step in the recovery of rods, if identified, may be a good candidate for the rate-limiting step in cones. This can be tested, for example, by examining the effect of overexpressing transducin GAP complex in cones. One should not observe accelerated recovery if transducin turnoff in cones is not rate-limiting. A breakthrough in single-cone recording techniques for mice in which approximately 3% of the photoreceptors are cones (Jeon et al. 1998) will provide the fullest harvest of all available genetic resources. Second, studies on the mechanism of reproducibility of the elementary responses of rods has a traditional focus on R* deactivation, which in the light of the new finding is a relatively fast step in recovery. Because one R* effectively activates hundreds of transducins during its brief lifetime, transduction gain at the interface of rhodopsin/transducin coupling and the rate at which transducin is deactivated must therefore contribute more than previously appreciated to the reproducibility of the elementary response of rods. Third but not the least, it is documented that the expression level of RGS in neurons changes in response to various experimental or pathologic conditions (Gold et al. 2002; Mirnics et al. 2001; Traynor and Neubig 2005). These changes may thus effectively affect the duration of the physiological responses these RGS proteins have over their cognate G protein signaling pathways in a variety of tissues. Adjusting the expression level of RGS proteins, e.g., by delivering extra copies of genes or by knocking it down using RNA interference through viral vectors, may thus be an efficient way of modulating neural activities.

Concluding remarks

Despite our knowledge of the basic biochemical mechanisms underlying the elementary response of rods, many important questions hereby identified remain to be satisfactorily answered. Areas that are particularly challenging include but are not limited to the mechanisms of reproducibility of a single-photon response, mechanisms of light/dark adaptation, rod/cone differences, and the structure/function relationship of known rod and cone photo-transduction genes under normal and pathologic conditions. This is still a fertile ground for insightful scientific discoveries.

References

Ala-Laurila P, Donner K, Koskelainen A (2004) Thermal activation and photoactivation of visual pigments. Biophys J 86:3653–3662
Angleson JK, Wensel TG (1993) A GTPase-accelerating factor for transducin, distinct from its effector cGMP phosphodiesterase, in rod outer segment membranes. Neuron 11:939–949
Aris L, Gilchrist A, Rens-Domiano S, Meyer C, Schatz PJ, Dratz EA, Hamm HE (2001) Structural requirements for the stabilization of metarhodopsin II by the C terminus of the alpha subunit of transducin. J Biol Chem 276:2333–2339
Arshavsky V, Bownds MD (1992) Regulation of deactivation of photoreceptor G-protein by its target enzyme and cGMP. Nature 357:416–417
Baehr W, Devlin MJ, Applebury ML (1979) Isolation and characterization of cGMP phosphodiesterase from bovine rod outer segments. J Biol Chem 254:11669–11677
Balasubramanian N, Levay K, Keren-Raifman T, Faurobert E, Slepak VZ (2001) Phosphorylation of the regulator of G-protein signaling RGS9-1 by protein kinase A is a potential mechanism of light- and Ca2+-mediated regulation of G-protein function in photoreceptors. Biochemistry 40:12619–12627
Barlow RB, Birge RR, Kaplan E, Tallent JR (1993) On the molecular origin of photoreceptor noise. Nature 366:64–66
Baumann C (1976) The formation of metarhodospin380 in the retinal rods of the frog. J Physiol 259:357–366
Baylor D (1996) How photons start vision. Proc Natl Acad Sci U S A 93:560–565
Baylor DA (1987) Photoreceptor signals and vision. Proctor lecture. Invest Ophthalmol Vis Sci 28:34–49
Baylor DA, Nunn BJ, Schnapf JL (1984) The photocurrent, noise and spectral sensitivity of rods of the monkey Macaca fascicularis. J Physiol 357:575–607
Birge RR, Knox BE (2003) Perspectives on the counterion switch-induced photoactivation of the G-protein-coupled receptor rhodopsin. Proc Natl Acad Sci U S A 100:9105–9107
Bortoff A (1964) Localization of slow potential responses in the Necturus retina. Vision Res 4:627–635
Bourne HR, Stryer L (1992) G-proteins. The target sets the tempo. Nature 358:541–543

Bulenger S, Marullo S, Bouvier M (2005) Emerging role of homo- and heterodimerization in G-protein-coupled receptor biosynthesis and maturation. Trends Pharmacol Sci 26:131–137

Burns ME, Baylor DA (2001) Activation, deactivation, and adaptation in vertebrate photoreceptor cells. Annu Rev Neurosci 24:779–805

Burns ME, Mendez A, Chen J, Baylor DA (2002) Dynamics of cyclic GMP synthesis in retinal rods. Neuron 36:81–91

Cai K, Itoh Y, Khorana HG (2001) Mapping of contact sites in complex formation between transducin and light-activated rhodopsin by covalent crosslinking: use of a photoactivatable reagent. Proc Natl Acad Sci U S A 98:4877–4882

Calvert PD, Makino CL (2002) The time course of light adaptation in vertebrate retinal rods. Adv Exp Med Biol 514:37–60

Calvert PD, Govardovskii VI, Krasnoperova N, Anderson RE, Lem J, Makino CL (2001) Membrane protein diffusion sets the speed of rod phototransduction. Nature 411:90–94

Chabre M, Cone R, Saibil H (2003) Biophysics: is rhodopsin dimeric in native retinal rods? Nature 426:30–31; discussion 31

Chen CK (2002) Recoverin and rhodopsin kinase. Adv Exp Med Biol 514:101–107

Chen CK, Inglese J, Lefkowitz RJ, Hurley JB (1995a) Ca(2+)-dependent interaction of recoverin with rhodopsin kinase. J Biol Chem 270:18060–18066

Chen CK, Burns ME, Spencer M, Niemi GA, Chen J, Hurley JB, Baylor DA, Simon MI (1999) Abnormal photoresponses and light-induced apoptosis in rods lacking rhodopsin kinase. Proc Natl Acad Sci U S A 96:3718–3722

Chen CK, Burns ME, He W, Wensel TG, Baylor DA, Simon MI (2000) Slowed recovery of rod photoresponse in mice lacking the GTPase accelerating-protein RGS9-1. Nature 403:557–560

Chen CK, Zhang K, Church-Kopish J, Huang W, Zhang H, Chen YJ, Frederick JM, Baehr W (2001) Characterization of human GRK7 as a potential cone opsin kinase. Mol Vis 7:305–313

Chen CK, Eversole-Cire P, Zhang H, Mancino V, Chen YJ, He W, Wensel TG, Simon MI (2003) Instability of GGL domain-containing RGS proteins in mice lacking the G-protein beta-subunit Gbeta5. Proc Natl Acad Sci U S A 100:6604–6609

Chen J, Makino CL, Peachey NS, Baylor DA, Simon MI (1995b) Mechanisms of rhodopsin inactivation in vivo as revealed by a COOH-terminal truncation mutant. Science 267:374–377

Chen TY, Peng YW, Dhallan RS, Ahamed B, Reed RR, Yau KW (1993) A new subunit of the cyclic nucleotide-gated cation channel in retinal rods. Nature 362:764–767

Cideciyan AV, Zhao X, Nielsen L, Khani SC, Jacobson SG, Palczewski K (1998) Null mutation in the rhodopsin kinase gene slows recovery kinetics of rod and cone phototransduction in man. Proc Natl Acad Sci U S A 95:328–333

Cone RA (1972) Rotational diffusion of rhodopsin in the visual receptor membrane. Nat New Biol 236:39–43

Cook NJ, Molday LL, Reid D, Kaupp UB, Molday RS (1989) The cGMP-gated channel of bovine rod photoreceptors is localized exclusively in the plasma membrane. J Biol Chem 264:6996–6999

Cooper A (1979) Energy uptake in the first step of visual excitation. Nature 282:531–533

Cote RH (2004) Characteristics of photoreceptor PDE (PDE6): similarities and differences to PDE5. Int J Impot Res 16 Suppl 1:S28–33

Cowan CW, Fariss RN, Sokal I, Palczewski K, Wensel TG (1998) High expression levels in cones of RGS9, the predominant GTPase accelerating-protein of rods. Proc Natl Acad Sci U S A 95:5351–5356

Craft CM, Whitmore DH (1995) The arrestin superfamily: cone arrestins are a fourth family. FEBS Lett 362:247–255

Crescitelli F (1984) The gecko visual pigment: the dark exchange of chromophore. Vision Res 24:1551–1553

Dizhoor AM, Ray S, Kumar S, Niemi G, Spencer M, Brolley D, Walsh KA, Philipov PP, Hurley JB, Stryer L (1991) Recoverin: a calcium sensitive activator of retinal rod guanylate cyclase. Science 251:915–918

Dizhoor AM, Olshevskaya EV, Henzel WJ, Wong SC, Stults JT, Ankoudinova I, Hurley JB (1995) Cloning, sequencing, and expression of a 24-kDa Ca(2+)-binding-protein activating photoreceptor guanylyl cyclase. J Biol Chem 270:25200–25206

Dryja TP (2000) Molecular genetics of Oguchi disease, fundus albipunctatus, and other forms of stationary night blindness: LVII Edward Jackson Memorial Lecture. Am J Ophthalmol 130:547–563

Dukkipati A, Kusnetzow A, Babu KR, Ramos L, Singh D, Knox BE, Birge RR (2002) Phototransduction by vertebrate ultraviolet visual pigments: protonation of the retinylidene Schiff base following photobleaching. Biochemistry 41:9842–9851

Ebrey TG (2000) pKa of the protonated Schiff base of visual pigments. Methods Enzymol 315:196–207

Edwards PC, Li J, Burghammer M, McDowell JH, Villa C, Hargrave PA, Schertler GF (2004) Crystals of native and modified bovine rhodopsins and their heavy atom derivatives. J Mol Biol 343:1439–1450

Erickson MA, Lagnado L, Zozulya S, Neubert TA, Stryer L, Baylor DA (1998) The effect of recombinant recoverin on the photoresponse of truncated rod photoreceptors. Proc Natl Acad Sci U S A 95:6474–6479

Fesenko EE, Kolesnikov SS, Lyubarsky AL (1985) Induction by cyclic GMP of cationic conductance in plasma membrane of retinal rod outer segment. Nature 313:310–313

Field GD, Rieke F (2002) Mechanisms regulating variability of the single photon responses of mammalian rod photoreceptors. Neuron 35:733–747

Fotiadis D, Liang Y, Filipek S, Saperstein DA, Engel A, Palczewski K (2003) Atomic-force microscopy: rhodopsin dimers in native disc membranes. Nature 421:127–128

Fuchs S, Nakazawa M, Maw M, Tamai M, Oguchi Y, Gal A (1995) A homozygous 1-base pair deletion in the arrestin gene is a frequent cause of Oguchi disease in Japanese. Nat Genet 10:360–362

Fung BK, Hurley JB, Stryer L (1981) Flow of information in the light-triggered cyclic nucleotide cascade of vision. Proc Natl Acad Sci U S A 78:152–156

Gold SJ, Heifets BD, Pudiak CM, Potts BW, Nestler EJ (2002) Regulation of regulators of G-protein signaling mRNA expression in rat brain by acute and chronic electroconvulsive seizures. J Neurochem 82:828–838

Gray-Keller MP, Detwiler PB (1994) The calcium feedback signal in the phototransduction cascade of vertebrate rods. Neuron 13:849–861

Gray-Keller MP, Polans AS, Palczewski K, Detwiler PB (1993) The effect of recoverin-like calcium-bindingproteins on the photoresponse of retinal rods. Neuron 10:523–531

Hagins WA, Yoshikami S (1974) Proceedings: a role for Ca2+ in excitation of retinal rods and cones. Exp Eye Res 18:299–305

Hardie RC (2001) Phototransduction in Drosophila melanogaster. J Exp Biol 204:3403–3409

He W, Cowan CW, Wensel TG (1998) RGS9, a GTPase accelerator for phototransduction. Neuron 20:95–102

Hecht SS, Pirenne MH (1942) Energy, quanta, and vision. J Gen Physiol 25:819–840

Horner TJ, Osawa S, Schaller MD, Weiss ER (2005) Phosphorylation of GRK1 and GRK7 by cAMP-dependent protein kinase attenuates their enzymatic activities. J Biol Chem 280:28241–28250

Hsu YT, Molday RS (1993) Modulation of the cGMP-gated channel of rod photoreceptor cells by calmodulin. Nature 361:76–79

Hu G, Wensel TG (2002) R9AP, a membrane anchor for the photoreceptor GTPase accelerating-protein, RGS9-1. Proc Natl Acad Sci U S A 99:9755–9760

Hu G, Jang GF, Cowan CW, Wensel TG, Palczewski K (2001) Phosphorylation of RGS9-1 by an endogenous protein kinase in rod outer segments. J Biol Chem 276:22287–22295

Hubbell WL, Cafiso DS, Altenbach C (2000) Identifying conformational changes with site-directed spin labeling. Nat Struct Biol 7:735–739

Hubbell WL, Altenbach C, Hubbell CM, Khorana HG (2003) Rhodopsin structure, dynamics, and activation: a perspective from crystallography, site-directed spin labeling, sulfhydryl reactivity, and disulfide cross-linking. Adv Protein Chem 63:243–290

Hurley JB, Chen J (2001) Evaluation of the contributions of recoverin and GCAPs to rod photoreceptor light adaptation and recovery to the dark state. Prog Brain Res 131:395–405

Hurley JB, Stryer L (1982) Purification and characterization of the gamma regulatory subunit of the cyclic GMP phosphodiesterase from retinal rod outer segments. J Biol Chem 257:11094–11099

Hurley JB, Dizhoor AM, Ray S, Stryer L (1993) Recoverin's role: conclusion withdrawn. Science 260:740

Imai H, Terakita A, Tachibanaki S, Imamoto Y, Yoshizawa T, Shichida Y (1997) Photochemical and biochemical properties of chicken blue-sensitive cone visual pigment. Biochemistry 36:12773–12779

Jastrzebska B, Maeda T, Zhu L, Fotiadis D, Filipek S, Engel A, Stenkamp RE, Palczewski K (2004) Functional characterization of rhodopsin monomers and dimers in detergents. J Biol Chem 279:54663–54675

Jeon CJ, Strettoi E, Masland RH (1998) The major cell populations of the mouse retina. J Neurosci 18:8936–8946

Johnson S, Michaelides M, Aligianis IA, Ainsworth JR, Mollon JD, Maher ER, Moore AT, Hunt DM (2004) Achromatopsia caused by novel mutations in both CNGA3 and CNGB3. J Med Genet 41:e20

Kang K, Bauer PJ, Kinjo TG, Szerencsei RT, Bonigk W, Winkfein RJ, Schnetkamp PP (2003) Assembly of retinal rod or cone Na(+)/Ca(2+)-K(+) exchanger oligomers with cGMP-gated channel subunits as probed with heterologously expressed cDNAs. Biochemistry 42:4593–4600

Kaupp UB, Hanke W, Simmoteit R, Luhring H (1988) Electrical and biochemical properties of the cGMP-gated cation channel from rod photoreceptors. Cold Spring Harb Symp Quant Biol 53:407–415

Kaupp UB, Niidome T, Tanabe T, Terada S, Bonigk W, Stuhmer W, Cook NJ, Kangawa K, Matsuo H, Hirose T et al. (1989) Primary structure and functional expression from complementary DNA of the rod photoreceptor cyclic GMP-gated channel. Nature 342:762–766

Kawamura S (1993) Rhodopsin phosphorylation as a mechanism of cyclic GMP phosphodiesterase regulation by S-modulin. Nature 362:855–857

Kefalov V, Fu Y, Marsh-Armstrong N, Yau KW (2003) Role of visual pigment properties in rod and cone phototransduction. Nature 425:526–531

Kefalov VJ, Estevez ME, Kono M, Goletz PW, Crouch RK, Cornwall MC, Yau KW (2005) Breaking the covalent bond—a pigment property that contributes to desensitization in cones. Neuron 46:879–890

Keresztes G, Martemyanov KA, Krispel CM, Mutai H, Yoo PJ, Maison SF, Burns ME, Arshavsky VY, Heller S (2004) Absence of the RGS9.Gbeta5 GTPase-activating complex in photoreceptors of the R9AP knockout mouse. J Biol Chem 279:1581–1584

Klein-Seetharaman J, Hwa J, Cai K, Altenbach C, Hubbell WL, Khorana HG (1999) Single-cysteine substitution mutants at amino acid positions 55–75, the sequence connecting the cytoplasmic ends of helices I and II in rhodopsin: reactivity of the sulfhydryl groups and their derivatives identifies a tertiary structure that changes upon light-activation. Biochemistry 38:7938–7944

Klenchin VA, Calvert PD, Bownds MD (1995) Inhibition of rhodopsin kinase by recoverin. Further evidence for a negative feedback system in phototransduction. J Biol Chem 270:16147–16152

Koch KW, Stryer L (1988) Highly cooperative feedback control of retinal rod guanylate cyclase by calcium ions. Nature 334:64–66

Korschen HG, Illing M, Seifert R, Sesti F, Williams A, Gotzes S, Colville C, Muller F, Dose A, Godde M et al. (1995) A 240 kDa protein represents the complete beta subunit of the cyclic nucleotide-gated channel from rod photoreceptor. Neuron 15:627–636

Koutalos Y, Nakatani K, Tamura T, Yau KW (1995a) Characterization of guanylate cyclase activity in single retinal rod outer segments. J Gen Physiol 106:863–890

Koutalos Y, Nakatani K, Yau KW (1995b) The cGMP-phosphodiesterase and its contribution to sensitivity regulation in retinal rods. J Gen Physiol 106:891–921

Krispel CM, Chen CK, Simon MI, Burns ME (2003a) Novel form of adaptation in mouse retinal rods speeds recovery of phototransduction. J Gen Physiol 122:703–712

Krispel CM, Chen CK, Simon MI, Burns ME (2003b) Prolonged photoresponses and defective adaptation in rods of Gbeta5$^{-/-}$ mice. J Neurosci 23:6965–6971

Kusnetzow AK, Dukkipati A, Babu KR, Ramos L, Knox BE, Birge RR (2004) Vertebrate ultraviolet visual pigments: protonation of the retinylidene Schiff base and a counterion switch during photoactivation. Proc Natl Acad Sci U S A 101:941–946

Lagnado L, Baylor D (1992) Signal flow in visual transduction. Neuron 8:995–1002

Lamb TD, Baylor DA, Yau KW (1979) The membrane current of single rod outer segments. Vision Res 19:385

Lambrecht HG, Koch KW (1991) A 26 kd calcium binding-protein from bovine rod outer segments as modulator of photoreceptor guanylate cyclase. EMBO J 10:793–798

Lambright DG, Sondek J, Bohm A, Skiba NP, Hamm HE, Sigler PB (1996) The 2.0 A crystal structure of a heterotrimeric G-protein. Nature 379:311–319

Langlois G, Chen CK, Palczewski K, Hurley JB, Vuong TM (1996) Responses of the phototransduction cascade to dim light. Proc Natl Acad Sci U S A 93:4677–4682

Lee RH, Navon SE, Brown BM, Fung BK, Lolley RN (1988) Characterization of a phosphodiesterase-immunoreactive polypeptide from rod photoreceptors of developing rd mouse retinas. Invest Ophthalmol Vis Sci 29:1021–1027

Leskov IB, Klenchin VA, Handy JW, Whitlock GG, Govardovskii VI, Bownds MD, Lamb TD, Pugh EN Jr, Arshavsky VY (2000) The gain of rod phototransduction: reconciliation of biochemical and electrophysiological measurements. Neuron 27:525–537

Lewis JW, Kliger DS (1992) Photointermediates of visual pigments. J Bioenerg Biomembr 24:201–210

Li TS, Volpp K, Applebury ML (1990) Bovine cone photoreceptor cGMP phosphodiesterase structure deduced from a cDNA clone. Proc Natl Acad Sci U S A 87:293–297

Liang J, Steinberg G, Livnah N, Sheves M, Ebrey TG, Tsuda M (1994) The pKa of the protonated Schiff bases of gecko cone and octopus visual pigments. Biophys J 67:848–854

Liang Y, Fotiadis D, Maeda T, Maeda A, Modzelewska A, Filipek S, Saperstein DA, Engel A, Palczewski K (2004) Rhodopsin signaling and organization in heterozygote rhodopsin knockout mice. J Biol Chem 279:48189–48196

Lin SW, Sakmar TP, Franke RR, Khorana HG, Mathies RA (1992) Resonance Raman microprobe spectroscopy of rhodopsin mutants: effect of substitutions in the third transmembrane helix. Biochemistry 31:5105–5111

Lyubarsky A, Nikonov S, Pugh EN Jr (1996) The kinetics of inactivation of the rod phototransduction cascade with constant Ca2+i. J Gen Physiol 107:19–34

Lyubarsky AL, Chen C, Simon MI, Pugh EN Jr (2000) Mice lacking G-protein receptor kinase 1 have profoundly slowed recovery of cone-driven retinal responses. J Neurosci 20:2209–2217

Lyubarsky AL, Naarendorp F, Zhang X, Wensel T, Simon MI, Pugh EN Jr (2001) RGS9-1 is required for normal inactivation of mouse cone phototransduction. Mol Vis 7:71–78

Makino CL, Dodd RL, Chen J, Burns ME, Roca A, Simon MI, Baylor DA (2004) Recoverin regulates light-dependent phosphodiesterase activity in retinal rods. J Gen Physiol 123:729–741

Makino ER, Handy JW, Li T, Arshavsky VY (1999) The GTPase activating factor for transducin in rod photoreceptors is the complex between RGS9 and type 5 G-protein beta subunit. Proc Natl Acad Sci U S A 96:1947–1952

Martemyanov KA, Lishko PV, Calero N, Keresztes G, Sokolov M, Strissel KJ, Leskov IB, Hopp JA, Kolesnikov AV, Chen CK, Lem J, Heller S, Burns ME, Arshavsky VY (2003) The DEP domain determines subcellular targeting of the GTPase activating protein RGS9 in vivo. J Neurosci 23:10175–10181

Matsumoto H, Tokunaga F, Yoshizawa T (1975) Accessibility of the iodopsin chromophore. Biochim Biophys Acta 404:300–308

McBee JK, Palczewski K, Baehr W, Pepperberg DR (2001) Confronting complexity: the interlink of phototransduction and retinoid metabolism in the vertebrate retina. Prog Retin Eye Res 20:469–529

Mendez A, Chen J (2002) Mouse models to study GCAP functions in intact photoreceptors. Adv Exp Med Biol 514:361–388

Mendez A, Burns ME, Roca A, Lem J, Wu LW, Simon MI, Baylor DA, Chen J (2000) Rapid and reproducible deactivation of rhodopsin requires multiple phosphorylation sites. Neuron 28:153–164

Mirnics K, Middleton FA, Stanwood GD, Lewis DA, Levitt P (2001) Disease-specific changes in regulator of G-protein signaling 4 (RGS4) expression in schizophrenia. Mol Psychiatry 6:293–301

Mou H, Cote RH (2001) The catalytic and GAF domains of the rod cGMP phosphodiesterase (PDE6) heterodimer are regulated by distinct regions of its inhibitory gamma subunit. J Biol Chem 276:27527–27534

Nakatani K, Yau KW (1988) Guanosine 3',5'-cyclic monophosphate-activated conductance studied in a truncated rod outer segment of the toad. J Physiol 395:731–753

Nakatani K, Chen C, Yau KW, Koutalos Y (2002) Calcium and phototransduction. Adv Exp Med Biol 514:1–20

Nathans J (1990) Determinants of visual pigment absorbance: identification of the retinylidene Schiff's base counterion in bovine rhodopsin. Biochemistry 29:9746–9752

Natochin M, Granovsky AE, Muradov KG, Artemyev NO (1999) Roles of the transducin alpha-subunit alpha4-helix/alpha4-beta6 loop in the receptor and effector interactions. J Biol Chem 274:7865–7869

Natochin M, Muradov KG, McEntaffer RL, Artemyev NO (2000) Rhodopsin recognition by mutant G(s)alpha containing C-terminal residues of transducin. J Biol Chem 275:2669–2675

Nishiguchi KM, Sandberg MA, Kooijman AC, Martemyanov KA, Pott JW, Hagstrom SA, Arshavsky VY, Berson EL, Dryja TP (2004) Defects in RGS9 or its anchor protein R9AP in patients with slow photoreceptor deactivation. Nature 427:75–78

Noel JP, Hamm HE, Sigler PB (1993) The 2.2 A crystal structure of transducin-alpha complexed with GTP gamma S. Nature 366:654–663

Osorio D, Nilsson DE (2004) Visual pigments: trading noise for fast recovery. Curr Biol 14:R1051–1053

Otto-Bruc AE, Fariss RN, Van Hooser JP, Palczewski K (1998) Phosphorylation of photolyzed rhodopsin is calcium-insensitive in retina permeabilized by alpha-toxin. Proc Natl Acad Sci U S A 95:15014–15019

Palczewski K, Subbaraya I, Gorczyca WA, Helekar BS, Ruiz CC, Ohguro H, Huang J, Zhao X, Crabb JW, Johnson RS et al. (1994) Molecular cloning and characterization of retinal photoreceptor guanylyl cyclase-activating-protein. Neuron 13:395–404

Palczewski K, Kumasaka T, Hori T, Behnke CA, Motoshima H, Fox BA, Le Trong I, Teller DC, Okada T, Stenkamp RE et al. (2000) Crystal structure of rhodopsin: A G-protein-coupled receptor. Science 289:739–745

Palczewski K, Sokal I, Baehr W (2004) Guanylate cyclase-activating-proteins: structure, function, and diversity. Biochem Biophys Res Commun 322:1123–1130

Park PS, Filipek S, Wells JW, Palczewski K (2004) Oligomerization of G-protein-coupled receptors: past, present, and future. Biochemistry 43:15643–15656

Pepperberg DR, Cornwall MC, Kahlert M, Hofmann KP, Jin J, Jones GJ, Ripps H (1992) Light-dependent delay in the falling phase of the retinal rod photoresponse. Vis Neurosci 8:9–18

Poo M, Cone RA (1974) Lateral diffusion of rhodopsin in the photoreceptor membrane. Nature 247:438–441

Preininger AM, Hamm HE (2004) G-protein signaling: insights from new structures. Sci STKE 2004:re3

Pugh EN Jr, Lamb TD (1993) Amplification and kinetics of the activation steps in phototransduction. Biochim Biophys Acta 1141:111–149

Qin N, Pittler SJ, Baehr W (1992) In vitro isoprenylation and membrane association of mouse rod photoreceptor cGMP phosphodiesterase alpha and beta subunits expressed in bacteria. J Biol Chem 267:8458–8463

　　　　　　　　Rev Physiol Biochem Pharmacol (2005) 155:101–121

Rahman Z, Gold SJ, Potenza MN, Cowan CW, Ni YG, He W, Wensel TG, Nestler EJ (1999) Cloning and characterization of RGS9-2: a striatal-enriched alternatively spliced product of the RGS9 gene. J Neurosci 19:2016–2026

Ramdas L, Disher RM, Wensel TG (1991) Nucleotide exchange and cGMP phosphodiesterase activation by pertussis toxin inactivated transducin. Biochemistry 30:11637–11645

Rieke F, Baylor DA (1998) Origin of reproducibility in the responses of retinal rods to single photons. Biophys J 75:1836–1857

Rivolta C, Sharon D, DeAngelis MM, Dryja TP (2002) Retinitis pigmentosa and allied diseases: numerous diseases, genes, and inheritance patterns. Hum Mol Genet 11:1219–1227

Robinson PR, Cohen GB, Zhukovsky EA, Oprian DD (1992) Constitutively active mutants of rhodopsin. Neuron 9:719–725

Ross EM, Wilkie TM (2000) GTPase-activating-proteins for heterotrimeric G-proteins: regulators of G-protein signaling (RGS) and RGS-like proteins. Annu Rev Biochem 69:795–827

Saari JC, Garwin GG, Van Hooser JP, Palczewski K (1998) Reduction of all-trans-retinal limits regeneration of visual pigment in mice. Vision Res 38:1325–1333

Sagoo MS, Lagnado L (1997) G-protein deactivation is rate-limiting for shut-off of the phototransduction cascade. Nature 389:392–395

Sakmar TP, Franke RR, Khorana HG (1989) Glutamic acid-113 serves as the retinylidene Schiff base counterion in bovine rhodopsin. Proc Natl Acad Sci U S A 86:8309–8313

Sakmar TP, Franke RR, Khorana HG (1991) The role of the retinylidene Schiff base counterion in rhodopsin in determining wavelength absorbance and Schiff base pKa. Proc Natl Acad Sci U S A 88:3079–3083

Sampath AP, Baylor DA (2002) Molecular mechanism of spontaneous pigment activation in retinal cones. Biophys J 83:184–193

Sampath AP, Matthews HR, Cornwall MC, Bandarchi J, Fain GL (1999) Light-dependent changes in outer segment free-Ca2+ concentration in salamander cone photoreceptors. J Gen Physiol 113:267–277

Sampath AP, Strissel KJ, Elias R, Arshavsky VY, McGinnis JF, Chen J, Kawamura S, Rieke F, Hurley JB (2005) Recoverin improves rod-mediated vision by enhancing signal transmission in the mouse retina. Neuron 46:413–420

Schertler GF, Hargrave PA (1995) Projection structure of frog rhodopsin in two crystal forms. Proc Natl Acad Sci U S A 92:11578–11582

Schnapf JL, Nunn BJ, Meister M, Baylor DA (1990) Visual transduction in cones of the monkey Macaca fascicularis. J Physiol 427:681–713

Schoenlein RW, Peteanu LA, Mathies RA, Shank CV (1991) The first step in vision: femtosecond isomerization of rhodopsin. Science 254:412–415

Schwarzer A, Schauf H, Bauer PJ (2000) Binding of the cGMP-gated channel to the Na/Ca-K exchanger in rod photoreceptors. J Biol Chem 275:13448–13454

Shichida Y, Imai H, Imamoto Y, Fukada Y, Yoshizawa T (1994) Is chicken green-sensitive cone visual pigment a rhodopsin-like pigment? A comparative study of the molecular properties between chicken green and rhodopsin. Biochemistry 33:9040–9044

Slepak VZ, Artemyev NO, Zhu Y, Dumke CL, Sabacan L, Sondek J, Hamm HE, Bownds MD, Arshavsky VY (1995) An effector site that stimulates G-protein GTPase in photoreceptors. J Biol Chem 270:14319–14324

Smith WC, Milam AH, Dugger D, Arendt A, Hargrave PA, Palczewski K (1994) A splice variant of arrestin. Molecular cloning and localization in bovine retina. J Biol Chem 269:15407–15410

Sondek J, Lambright DG, Noel JP, Hamm HE, Sigler PB (1994) GTPase mechanism of Gproteins from the 1.7-A crystal structure of transducin alpha-GDP-AIF-4. Nature 372:276–279

Sprang SR (1997) G-protein mechanisms: insights from structural analysis. Annu Rev Biochem 66:639–678

Steinberg G, Ottolenghi M, Sheves M (1993) pKa of the protonated Schiff base of bovine rhodopsin. A study with artificial pigments. Biophys J 64:1499–1502

Struthers M, Yu H, Oprian DD (2000) G-protein-coupled receptor activation: analysis of a highly constrained, "straitjacketed" rhodopsin. Biochemistry 39:7938–7942

Stryer L (1991) Visual excitation and recovery. J Biol Chem 266:10711–10714

Suda K, Filipek S, Palczewski K, Engel A, Fotiadis D (2004) The supramolecular structure of the GPCR rhodopsin in solution and native disc membranes. Mol Membr Biol 21:435–446

Tachibanaki S, Arinobu D, Shimauchi-Matsukawa Y, Tsushima S, Kawamura S (2005) Highly effective phosphorylation by G-protein-coupled receptor kinase 7 of light-activated visual pigment in cones. Proc Natl Acad Sci U S A 102:9329–9334

Taylor WR, Smith RG (2004) Transmission of scotopic signals from the rod to rod-bipolar cell in the mammalian retina. Vision Res 44:3269–3276

Tomita T, Kaneko A, Murakami M, Pautler EL (1967) Spectral response curves of single cones in the carp. Vision Res 7:519–531

Travis GH (2005) DISCO! Dissociation of cone opsins: the fast and noisy life of cones explained. Neuron 46:840–842

Traynor JR, Neubig RR (2005) Regulators of G-protein signaling and drugs of abuse. Mol Interv 5:30–41

Tsang SH, Burns ME, Calvert PD, Gouras P, Baylor DA, Goff SP, Arshavsky VY (1998) Role for the target enzyme in deactivation of photoreceptor G-protein in vivo. Science 282:117–121

Vuong TM, Chabre M, Stryer L (1984) Millisecond activation of transducin in the cyclic nucleotide cascade of vision. Nature 311:659–661

Wassle H (2004) Parallel processing in the mammalian retina. Nat Rev Neurosci 5:747–757

Watson AJ, Katz A, Simon MI (1994) A fifth member of the mammalian G-protein beta-subunit family. Expression in brain and activation of the beta 2 isotype of phospholipase C. J Biol Chem 269:22150–22156

Watson AJ, Aragay AM, Slepak VZ, Simon MI (1996) A novel form of the G-protein beta subunit Gbeta5 is specifically expressed in the vertebrate retina. J Biol Chem 271:28154–28160

Weiss ER, Ducceschi MH, Horner TJ, Li A, Craft CM, Osawa S (2001) Species-specific differences in expression of G-protein-coupled receptor kinase (GRK) 7 and GRK1 in mammalian cone photoreceptor cells: implications for cone cell phototransduction. J Neurosci 21:9175–9184

Weitz D, Ficek N, Kremmer E, Bauer PJ, Kaupp UB (2002) Subunit stoichiometry of the CNG channel of rod photoreceptors. Neuron 36:881–889

Wilden U, Hall SW, Kuhn H (1986) Phosphodiesterase activation by photoexcited rhodopsin is quenched when rhodopsin is phosphorylated and binds the intrinsic 48-kDa protein of rod outer segments. Proc Natl Acad Sci U S A 83:1174–1178

Woodruff ML, Sampath AP, Matthews HR, Krasnoperova NV, Lem J, Fain GL (2002) Measurement of cytoplasmic calcium concentration in the rods of wild-type and transducin knock-out mice. J Physiol 542:843–854

Xu J, Dodd RL, Makino CL, Simon MI, Baylor DA, Chen J (1997) Prolonged photoresponses in transgenic mouse rods lacking arrestin. Nature 389:505–509

Yamamoto S, Sippel KC, Berson EL, Dryja TP (1997) Defects in the rhodopsin kinase gene in the Oguchi form of stationary night blindness. Nat Genet 15:175–178

Yamazaki M, Li N, Bondarenko VA, Yamazaki RK, Baehr W, Yamazaki A (2002) Binding of cGMP to GAF domains in amphibian rod photoreceptor cGMP phosphodiesterase (PDE). Identification of GAF domains in PDE alphabeta subunits and distinct domains in the PDE gamma subunit involved in stimulation of cGMP binding to GAF domains. J Biol Chem 277:40675–40686

Yau KW, Nakatani K (1985a) Light-induced reduction of cytoplasmic free calcium in retinal rod outer segment. Nature 313:579–582

Yau KW, Nakatani K (1985b) Light-suppressible, cyclic GMP-sensitive conductance in the plasma membrane of a truncated rod outer segment. Nature 317:252–255

Yau KW, Lamb TD, Matthews G, Baylor DA (1979a) Current fluctuations across single rod outer segments. Vision Res 19:387–390

Yau KW, Matthews G, Baylor DA (1979b) Thermal activation of the visual transduction mechanism in retinal rods. Nature 279:806–807

Zhang K, Howes KA, He W, Bronson JD, Pettenati MJ, Chen C, Palczewski K, Wensel TG, Baehr W (1999) Structure, alternative splicing, and expression of the human RGS9 gene. Gene 240:23–34

Zhang X, Wensel TG, Kraft TW (2003) GTPase regulators and photoresponses in cones of the eastern chipmunk. J Neurosci 23:1287–1297

Zhao X, Haeseleer F, Fariss RN, Huang J, Baehr W, Milam AH, Palczewski K (1997) Molecular cloning and localization of rhodopsin kinase in the mammalian pineal. Vis Neurosci 14:225–232

Zheng J, Trudeau MC, Zagotta WN (2002) Rod cyclic nucleotide-gated channels have a stoichiometry of three CNGA1 subunits and one CNGB1 subunit. Neuron 36:891–896

Zhong H, Molday LL, Molday RS, Yau KW (2002) The heteromeric cyclic nucleotide-gated channel adopts a 3A:1B stoichiometry. Nature 420:193–198

Zhu X, Brown B, Li A, Mears AJ, Swaroop A, Craft CM (2003) GRK1-dependent phosphorylation of S and M opsins and their binding to cone arrestin during cone phototransduction in the mouse retina. J Neurosci 23:6152–6160

Zhukovsky EA, Oprian DD (1989) Effect of carboxylic acid side chains on the absorption maximum of visual pigments. Science 246:928–930

Zimmerman AL (2002) Two B or not two B? Questioning the rotational symmetry of tetrameric ion channels. Neuron 36:997–999